Submarines, Diving, and the Underwater World

A Bibliography

Compiled and Edited by
Frank J. Anderson

ARCHON BOOKS
1975

Library of Congress Cataloging in Publication Data

Anderson, Frank J.
Submarines, diving, and the underwater world.

1. Submarine boats—Bibliography. 2. Submarine warfare—Bibliography. 3. Diving, Submarine—Bibliography. 4. Underwater exploration—Bibliography. I. Title.
Z6834.S9A5 1975 [VM365] 016.3593'2'57 75-16297
ISBN 0-208-01508-6

First published 1975 as an Archon Book,
an imprint of
The Shoe String Press, Inc.
Hamden, Connecticut 06514

Printed in the United States of America

Dedicated to those who venture under the sea

PREFACE

Back in 1963 my bibliography of underwater literature was published with the title of *Submarines, Submariners, and Submarining*. That listing contained 650 main entries arranged on a divided plan, with author entries and title entries in separate sections. The subjects were coded and the code numbers were part of the main entries with a coded list of subject entries following the primary listings. An appendix listed books in foreign languages. This arrangement was more complicated than it needs to be.

The increased number of listings, shift in subject matter, and new arrangement indicate that this current bibliography is a new work and warrants a new title. Thus this bibliography has been titled *Submarines, Diving, and the Underwater World*. The present bibliography includes the author, title entries, and subject cross references within a single alphabet. English language titles and foreign language titles are within this single alphabet. The author entry is the main entry and is a numbered entry. The main entries have full bibliographical information concerning the book, including Library of Congress catalog card numbers, and where available the International Standard Book Numbers (these numbers may at some future time afford access to information stored in a computer data bank). Title listings, subsidiary authors, and subject entries are cross referenced to the numbered main entry. This arrangement should facilitate the use of this bibliography.

The books which were listed in *Submarines, Submariners, and Submarining* dealt primarily with the submarine as a war machine. Since 1963 international interest in the oceans has greatly intensified and a variety of nonmilitary submarines have been at work underwater. The United States has passed laws creating Sea Grant colleges to encourage oceanographic research. These Sea Grant colleges are based on the same idea as that of the foundation of Land Grant colleges, a hundred years earlier, which were established for the encouragement of agricultural research.

Every nation with a seacoast has laid claim to miles and miles of the adjacent ocean, and to the seabed beneath it. The continental limits and sovereignty over the seabed have created such confusion and controversy that an international conference was held in Caracas in 1974 in an attempt to resolve some of the conflicting claims. The conference was indecisive except that the participants have agreed to meet again and have another conference on the matter. Some of the factors involved in this burgeoning interest in the seas are fisheries, petroleum, mineral resources, conservation, navigational access, and national security. Some nations are opposed to the idea of individual nations having legal claim to oceanic and seabed territory, contending that the oceans are a global resource in which all nations should have a share. The questions of territorial waters and the seabed claims are questions which will take years, and perhaps decades, to resolve.

Since 1963 the oceans have become much more than a submarine battlefield. Skin and scuba diving has been perfected as a sport and the oceans have become valuable as a recreational area. Diving enthusiasts, familiar with the beauties of the underwater world and aware of the fragility of that world, are conservationists who object to having their playgrounds despoiled with mining, dredging, and drilling operations. On the other hand there are commercial interests who believe that the sea would be more beautiful if it were covered with oil rigs. They bolster their arguments with cries about our national oil shortage. Underwater archaeology has advanced to the point where it is a disciplined science which is constantly adding to our store of historical information and which is drastically changing our ideas about the chronology of events in ancient history. The scientists are in the seas now, and these investigators are not limited to oceanographers. Just about every scientific discipline known has become involved in surveying the watery environment; not only to learn about ocean currents, salinity, and mineral content but also to investigate the seas as a place in which man can live and operate. Physicists, chemists, biologists, physiologists, archaeologists, and psychologists are all in on the action.

Submarines, diving, and the underwater world are the topics of the approximately 1500 books listed in this bibliography. The list includes histories, memoirs, fiction, technical works, biographies, bibliographies, and how-to-do-it manuals. There are explanatory and adventurous books for the young reader, as well as books in the *Up Scope! Down Scope! Fire One!* genre for the armchair underwater warrior. There are technical treatises and documentary data books for those who require objective and factual materials.

Official, quasi-official, and unofficial histories of the wars in which submarines have been employed are listed. These books come from the pens and typewriters of adversaries and allies, of victims and victors.

Works regarding the technological advances made in man's "mastery" of the sea are listed; such as accounts of bathyscaphe probes, underwater drift voyages, Sealab living, and deep diving with the use of the aqualung. The ocean's contempt of man's "mastery" is also recorded in the literature of submarine disasters and the subsequent rescue and salvage attempts.

The books in this list range over a period of more than 400 years of man's attempts to use the underwater world for his own ends. Approximately 285 of the books listed were published after 1964, and include a few titles published as late as 1975. Approximately 565 titles published before 1964 are new to this listing.

The books listed are written in various languages with English language books predominating. There is a good representation of books in French and in German. A few titles in Russian and Spanish are included, along with a token representation of Dutch, Hungarian, Japanese, Scandinavian, and Turkish language books.

Fifteen hundred books are quite a few books, but they do not comprise the sum total of everything written about submarines, diving, and the underwater world. In addition to books there is a vast profusion of papers in scientific and technical journals, plus the pages and pages of articles and stories which have appeared in popular magazines and newspapers. Bibliographical apparatus in the form of periodical and newspaper indexes exist to get at this literature, and the interested scholar can ferret out information. Add to the journals a tremendous body of report literature produced by naval and military agencies, manufacturing concerns, and university researchers in just about every country of the world. Consider also the operational log books of individual submarines, the war patrol reports, and the logs and reports of antisubmarine forces. Compound all of that with the knowledge that much of that material is not generally available, but kept secret or confidential due to a nation's security requirements.

Pile Pelion on Ossa with the thousands of patents which have been issued for submarines, diving equipment, and related things—again on an international basis—and one begins to see that some 1500 books are but the tip of a periscope which indicates the existence of the submerged hull.

It is rather doubtful if anyone could ever compile a registry of "all of the literature" relating to submarines and the underwater world. This bibliography is an attempt to list books related to underwater topics and to serve the scholar and researcher with a starting point. The compiler hopes that this work will be useful to students, researchers, librarians, book dealers and book collectors. The desire may have been to achieve perfection, but the only one who is perfect is Allah, and the compiler does not presume to compete with him. The compiler is aware that there are probably errors of commission and ommission within the following pages and accepts full responsibility for these errors.

Acknowledgements and Thanks

Submarines, Submariners and Submarining has been out-of-print for about five years, and during that time Frances Rutter, publisher of the Shoe String Press, has been inviting me to prepare a new bibliography covering an even broader area. Without her encouragement and cajolery I would never have committed myself to the task. Compiling, editing, and typing an extensive bibliography is sheer drudgery. I thank Mrs. Rutter for her confidence in my ability, but I hope that I have a computer available for the next up-date when *this* book goes out-of-print and she starts nagging me again.

My thanks to John H. Albrecht of Toronto, who is a student of submarine history and a collector of submarine books, for his interest, assistance, and encouragement.

Thanks to my friends, who haven't had any letters from me during the past few months, for still being my friends.

Thanks to my wife Jeanette and my daughter Maria for their loving forbearance which kept them from bugging me about the things which needed doing around the house while I was working on this bibliography.

The searching out of titles for inclusion in this list, and their bibliographical verification, has been accomplished primarily in the Sandor Teszler Library of Wofford College. Without the availability of that scholarly resource this work might have died aborning, or taken another ten years to complete.

Thanks to the power and light company for illumination to work by during the late night and early morning hours.

Thanks to the librarians, book dealers, and collectors for causing *Submarines, Submariners, and Submarining* to go out-of-print, and for suggesting this new version be published.

And finally, thank the Lord this book is done.

—FRANK J. ANDERSON

Spartanburg, South Carolina

A

1 AALL, Herman Johan Regnor Harris, 1871- . *Undervandskrigen og Verdensdespotiet.* Kristiania, Parmann, 1918. 264 pp.

Abandon Ship! See entry 912.

Abbot, Frederic Vaughn, 1858- . See entry 1099.

2 ABBOT, Henry Larcom, 1831-1927. *The Beginning of Modern Submarine Warfare, Under Captain-Lieutenant David Bushnell, Sappers and Miners, Army of the Revolution. Being a Historical Compilation. . . .* Willets Point, N.Y.H., Printed on the Battalion Press, Sergt. Carmichael and Pvt. Beck, printers, 1881. (U.S. Engineer School. Professional Papers Vol. I, No. 3) 34 pp., illus. LC 13-11682.

Has running title "Bushnell's Submarine Warfare."

An important source book which has been used by all subsequent writers on American submarine history.

3 ————. *Beginning of Modern Submarine Warfare Under Captain Lieutenant David Bushnell.* (Archon Books) Hamden, Shoe String Press, 1966. 69 pp., illus. LC 66-17740.

A facsimile reproduction of the rare 1881 pamphlet, with biographical appendixes and a bibliography by Frank J. Anderson.

4 ————. *Course of Lectures Upon the Defence of*

the Sea-Coast of the United States, Delivered before the U.S. Naval War College . . . 1887. N.Y., Van Nostrand, 1888. 167 pp. LC 3-498C.

5 ________. *Notes on Submarine Mining in England, in the Autumn of 1883.* Willets Point, N.Y.H., Printed on the Battalion Press, Sergt. Carmichael, Corp. Beck and Pvt. Phillips, printers, 1883? (U.S. Engineer School. Paper no. VII) 8 p., illus. LC 13-11679.

6 ________. *Report Upon Experiments and Investigations to Develop a System of Submarine Mines for Defending the Harbors of the United States.* Washington, Government Printing Office, 1881. (Professional Papers of the Corps of Engineers, U.S. Army. No. 23) 444 pp., illus., tables. LC 12-16142 rev 2.

7 ABBOT, Willis John, 1863-1934. *Aircraft and Submarines: the Story of the Invention, Development, and Present-Day Uses of War's Newest Weapons.* N.Y. and London, Putnam, 1918. 388 pp., illus. LC 18-9304.

Die Abenteurliche Fahrt der Nautilus. See entry 18.

8 ABKOWITZ, Martin A. *Stability and Motion Control of Ocean Vehicles.* (NSF / Sea Grant Project GH-1). Cambridge, Massachusetts Institute of Technology, 1969. 190 pp. + 154 pp. appendices. Bibliography, diagrams, formulae. LC 70-93041. SBN 262-51006-5 (pbk).

Aboukir, HMS See entry 1441.

About Submarines. See entry 188.

Above and Below. See entry 1123.

Above Us the Waves. See entry 1425.

Ackerman, Jack. See entry 712.

9 ACKWORTH, Bernard, 1885- . *Life in a Submarine.* London, Tuck, 1941. 48 pp., illus. LC 42-22115.

Ackworth retired from the Royal Navy as a Captain. Won the DSO in 1917. Served 18 years in submarines. Served as commander

Acoustic Triangulation . . . See entry 887.

Acoustics. See entry 662.

Action in Submarines. See entry 1458.

of ASW forces during four years of WW II. Closely associated with the development of ASDIC. See also *Who's Who, 1959*.

Adamson, Hans Christian, 1890- . See entries 789, 790, 793, 794.

Admiral Rickover and the Nuclear Navy. See entry 287.

The Admiral's Wolf Pack. See entry 923.

The Admiralty Regrets. See entry 1426.

The Adventures of the Piccards. See entry 622.

The Adventures of the U-202. See entry 1156.

Adventures Under the Sea. See entry 1244.

An Aerial Race . . . with Brief References to a Submarine. . . . See entry 180.

Affray, HMS. See entry 537.

10 AIGNER, Franz, 1882- . *Unterwasserschalltechnik . . . Grundlagen, Ziele und Grenzen.* Berlin, Krayn, 1922. 322 pp., illus. LC P026-1.

Aircraft and Submarines. See entry 7.

Aircraft Versus Submarine. See entry 997.

11 AJAX, pseud. *The German Pirate: His Methods and Record.* London, Pearson, 1918. 120 pp. LC 24-21755.

"Accounts of German submarine exploits . . . compiled from British Admiralty Documents and Sworn Statements of Survivors." Foreword.

Alarm! Tauchen!! U-Boot in Kampf und Sturm. See entry 460.

Albers, V. M. See entry 662.

Algae and Submarine Habitability. . . . See entry 1387.

Alger, Philip R. See entry 285a.

Allen, Ian. See entry 526.

12 ALMAN, Karl. *Angriff, Ran, Versenken.* Rastatt, Pabel Verlag, 1965. 326 pp., illus.

13 ————. *Ritter der Sieben Meere.* Rastatt, Pabel Verlag, 1963. 304 pp., illus., tables.

Accounts of WW II submarine combat.

Als Führer der U-Boote in Weltkriege. See entry 65.

Als U-Boots-Kommandant Gegen England. See entry 416.

Alves, D. See entry 1350.

Alvin (research submarine). See entries 146, 403.

Alvis, Harry J. See entries 1338, 1339, 1346.

Amazing Adventure. See entry 213.

Amerdroz, R. T. See entry 331.

14 AMERICAN BUREAU OF SHIPPING. *Guide for the Classification of Manned Submersibles.* N.Y., American Bureau of Shipping, 1968. 47 pp.

Technical work on classification, standards, and certification of commercial and research submersibles.

15 AMERICAN INSTITUTE OF AERONAUTICS AND ASTRONAUTICS. *Symposium on Modern Developments in Marine Sciences.* (Monograph Series, Vol. 4) North Hollywood, CA, Western Periodicals, 1971.

Midget submarines, commercial and scientific.

The American Navy in Europe. See entry 763.

American Revolution. See entries 2, 3, 51.

American Ship Casualties of the World War. . . . See entry 1384.

American Submarines. See entries 770, 771.

American Submarines. See entries 7, 14, 46, 50, 52, 53, 71-76, 91, 96, 175, 179, 199, 208, 209, 235, 243, 248, 249, 251, 252, 253, 281, 296, 304, 305, 314, 352, 361, 397, 399, 429, 564-568, 606, 617, 618, 621, 625, 633, 652, 653, 654, 731, 732, 759, 762, 763, 769, 770, 771, 788-794, 798, 808, 822, 823, 950, 957, 988, 989, 992, 1069, 1070, 1071, 1095, 1126, 1127, 1137, 1138, 1144, 1169, 1170, 1173, 1175, 1185, 1227, 1239, 1240, 1241, 1248, 1255, 1266, 1268.

Amerikanische Menschlichkeit im Lichte des Diplomatischen Notenwechsels. See entry 173.

An Bord. See entry 400.

16 ANDERSON, Frank J. *Submarines, Submariners, Submarining: A Checklist of Submarine Books in the English Language, Principally of the Twentieth Century.* Hamden, Shoe String Press, 1963. 140 pp. LC 63-22083.

A bibliography arranged by author, subject and title. With an appendix listing foreign language titles. Approximately 650 titles cited.

Anderson, Frank J. See entry 3.

17 ANDERSON, Jane. *Flying, Submarining and Mine Sweeping.* London, Causton, 1916. 36 pp. LC 16-23833.

Newspaper type "feature story" writing.

18 ANDERSON, William Robert, 1921- with Clay Blair. *Die Abenteurliche Fahrt der Nautilus.* Munich, Desch, 1959. 198 pp., illus.

German translation of *Nautilus 90° North.*

19 ANDERSON, William Robert, 1921- and others. *Atomic Submarines.* Front matter illustrated by Robert Borja. (New ed.) Chicago, Children's Press, 1968. 63 pp., illus. LC 68-14726.

20 ANDERSON, William Robert, 1921- . *First Under the North Pole: The Voyage of the Nautilus.* Cleveland, World, 1959. 64 pp., illus., maps. LC 59-7745.

Author was captain of the *Nautilus* when he took the world's first nuclear powered submarine under the North Pole.

21 ANDERSON, William Robert, 1921- and Clay Blair. *Nautilus 90° Nord.* Paris, Arthaud, 1959.

22 ANDERSON, William Robert, 1921- with Clay Blair, Jr. *Nautilus 90° North.* Cleveland, World, 1959. 251 pp. Illus. with photographs by John Krawczyk. Map endpapers. LC 59-6656.

Lists personnel aboard during Nautilus transpolar voyage of 1958.

23 ________. *Nautilus 90° North.* (Signet Book D1692) N.Y., New American Library, 1959. 144 pp., illus. with photographs by John Krawczyk.

Anderson, William Robert, 1921- . See entry 1438.

24 ANDREWS, Graeme. *Fighting Ships of Australia and New Zealand, 1973-1974 Edition.* Kogarah Bay, Regency House, 1973. 64 pp., illus.

Originally published in 1971 with the title *Australasian Navies.*

Andrews, Philip. See entry 356.

L'Angoisse des Veilles Sousmarines. See entry 543.

Angriff, Ran, Versenken. See entry 12.

An Annotated Bibliography of Submarine Technical Literature, 1557 to 1953. See entry 894.

An Annotated Literature Survey of Submarines. . . . See entry 157.

25 ANSCOMBE, Charles. *Submariner.* Foreword by Sir Sidney Raw. London, Kimber, 1957. 203 pp., plates. LC 58-26263.

Anti-Submarine Warfare. See entries 37, 160, 182-185, 207, 214, 216-219, 246, 331, 401, 486, 608, 931, 977, 997, 1012, 1013, 1015, 1053, 1104, 1171, 1381, 1469.

26 ANTIER, Jean-Jacques. *Histoire Mondiale du Sous-Marin.* Paris, Laffont, 1968. 389 pp., illus., tables.

L'Appel des Profondeurs. See entry 280.

27 APPLETON, Victor (pseud.) *Tom Swift and His Submarine Boat: Or, Under the Ocean for Sunken Treasure.* N.Y., Grosset & Dunlap, 1910. 216 pp., frontis. Juvenile fiction.

28 APPLETON, Victor, 2nd (pseud.). *Tom Swift and His Jetmarine.* (The New Tom Swift, Jr. Adventures) illustrated by Graham Kaye. N.Y., Grosset, 1954: London, Low, 1960. LC 54-372.

29 ________. *Tom Swift and His Subocean Geotron.* Illus. by Edward Moritz. London, Collins, 1969. 160 pp., illus. SBN 00-162211-0.

L'Apogée de la Guerre Sous-Marine. See entry 1067.

The Aquanauts. See entry 1093.

Archaeology Under Water. See entry 58.

30 ARCHER, William, 1856-1924. *The Pirate's Progress: A Short History of the U-Boat.* London, Chatto & Windus, 1918. 96 pp. LC 18-17007.

31 ________. *Sφrφverens faerd.* Copenhagen, Pio, 1918. 84 pp. LC 24-21756.

An authorized Danish and Norwegian edition translated by Johan Warburg.

William Archer was born in Perth, Scotland and educated at Edinburgh University. Scottish drama critic and playwright, trained as a lawyer. Drama critic for London *Figaro,* 1879-81. Introduced Ibsen to the English-speaking public. At outbreak of war in 1914 he devoted himself to writing propaganda for the British government. His only son was killed in the war in 1918. He turned to spiritualism and wrote on the topic. See *Dictionary of National Biography*.

Arctic. See entry 1463.

Arctic Submarine. See entry 831.

Argentine Navy. See entry 746.

Argonaut. See entry 742.

Armament Conference Treaties. See entry 1431.

Arms Limitation. See entries 186, 1140, 1399, 1431-34, 1436, 1437.

32 ARMSTRONG, Sir George Elliot, 1866- . *Torpedoes and Torpedo-Vessels.* (Royal Navy Handbook) London, Bell, 1896. xvi, 287 pp., frontis. by Chevalier E. de Martino. illus., diagrams. LC 13-5896.

A second edition was published in 1901.

Arnauld de la Perière, Lotharvon. See entry 601.

Around the World Submerged: The Voyage of the Triton. See entry 71.

The Art of the Aqualung. See entry 541.

Art of Underwater Photography. See entry 1172.

Assault on a Queen. See entry 406.

ASW: Anti-Submarine Warfare. See entry 221.

ASW, Anti-Submarine Warfare. See entry 1381.

Athenia (steamship). See entry 203.

33 ATHERLY-JONES, Llewellyn Archer, 1851- . *The Deutschland.* (In Grotius Society. *Problems of the War. . .*

London, 1916-19. Vol. III, 1918, p. 37-41.) LC 21-1020.
Concerning the merchant submarine Deutschland.

Atlantic Torpedo. See entry 583.

Atomic Submarine. See entry 833.

The Atomic Submarine. See entry 118.

The Atomic Submarine. See entry 613.

The Atomic Submarine. See entry 776.

The Atomic Submarine and Admiral Rickover. See entries 117, 118.

Atomic Submarines. See entry 988.

Atomic Submarines. See entries 18-23, 38, 71, 75, 76, 96, 122, 131, 174, 179, 248, 249, 257, 361, 385, 489, 560, 570, 606, 776, 989, 1059, 1095, 1170, 1173, 1174, 1175, 1248, 1279, 1281, 1282, 1388, 1391, 1438.

Atomnye Podvodnye Lodki. See entry 174.

. . . Atmospheric Purification . . . on Nuclear-Powered Submarines. See entries 1388, 1392.

The Attack Submarine. See entry 731.

34 AUBOSSON, Marquis de la Feuillade d'. *Memoire sur les Bateaux Sous-Marin.* Paris (?), 1840.

35 AUER, James E. *The Postwar Rearmament of Japanese Maritime Forces 1945-71.* N.Y., Praeger, 1973. 345 pp., illus. LC 72-83564 ISBN 0-275-28633-9.

Auf Gefechtsstationen! See entry 562.

Auftauchen! Kriegsfahrten von U-53. See entry 1072.

Augen Durch Nacht und Nebel. See entry 85.

Auguste Piccard, Captain of Space, Admiral of the Abyss. See entry 402.

Auld, Frank, Jr. See entry 1340.

Aus dem Logbuch des I. Wachoffiziers. U-66. See entry 727.

Ausgelaufen Westwärts. See entry 1011.

Die Aushungerung Englands. See entry 1113.

Australian Navy. See entry 24.

Austrian Submarines. See entry 1261.

36 AUSRO-HUNGARIAN MONARCHY. Reichs-kriegs-ministerium. Marine Section. *Unterricht im Taucherdienst in der K. K. Kriegs-Marine. . . .* Vienna, K. K. Hof-und Staatsdruckerei, 1880-85. 2 volumes. Folding plates and tables. LC 9-5080.

Deals with equipment and techniques of deep-sea diving.

Austro-Hungarian Navy. See entry 1147.

Austro-Hungarian Submarines. See entry 728.

37 AUTEN, Harold, 1891- . *Q Boat Adventures: The Exploits of the Famous Mystery Ships.* (First ed.) London, Jenkins, 1919. 280 pp., frontispiece, ports, illus.

"Q" ships were a British anti-submarine measure during World War One. The author was a Q boat commander.

L'Aventure Sous-Marine. See entries 308, 312.

L'Aviation Sous-Marine. See entry 1016.

Axis Submarine Manual. See entry 1379.

B

38 BAAR, James, 1929- and William E. HOWARD. *Polaris!* N.Y., Harcourt, 1960. 245 pp., plates. (W) 60-12731.

About the Polaris missile system and submarines.

39 BACMEISTER, Walter. *Der U-Bootkrieg als Weg zum Endsieg.* Weimar, Duncker, 1917. 30 pp. LC 54-55976.

40 BACON, Francis, viscount St. Albans, 1561-1626. *Novum Organum Scientiarum.* London, 1620.

Has been reprinted many times since initial publication. Reference to the diving bell.

41 BACON, Reginald Hugh Spencer, 1863-1947. *From 1900 Onward.* Foreword by Sir Archibald Hurd, and an introduction to the war chapters by Sir. E. R. G. R. Evans. London, Hutchinson, 1939: Toronto, Ryerson Press, 1939. 398 pp., plates, illus. LC A41-2659.

42 BACON, Roger, 1214?-1294. *De L'Admirable Pouvoir de L'art et de la Nature.* Lyon, 1557.

A rare book.

42a ________. *Roger Bacon's Letter Concerning the Marvelous Power of Art and of Nature and Concerning the Nullity of Magic.* Translated from the Latin by Tenney L. Davis . . . together with notes and an account of Bacon's life and work. Easton, Penna., The Chemical Publishing Co., 1923. 76 pp., facsimiles, portrait. LC 23-17153.

43 ________. *The Mirror of Alchemy.* 1597.

Discusses an instrument with which men could walk under the sea, no doubt a diving bell.

44 ________. *De Secretis Operibus Artis et Naturae.* Lyons, 1557. Translated by G. De Tournus.

45 BAGNASCO, Ermino. *I Sommergibili della Seconda Guerra Mondiale.* Parma, Ermano Albertelli, 1973. 336 pp., illus., diagrams.

Deals with the submarines of all of the nations involved in WW II. Includes cross-section diagrams of submarine classes.

Baldiger U-Boot-Friede? See entry 871.

46 BANNING, Kendall, 1879- . *Submarine! The Story of Undersea Fighters.* Illustrated by Charles Rosner. N.Y., Random House, 1942. 51 pp., illus. LC 42-24607.

47 BANNISTERE, Ducoux. *Notice sur Denis Papin.* Blois, 1854.

48 BARADA, Bill *Underwater: The Skindiver's Manual.* Los Angeles, Trend Books, 1955. 127 pp., illus. LC 55-3696.

Baralong, HMS. See entries 276, 277, 477, 479, 522.

Der Baralong-Fall. See entry 479.

Barb, USS (SS-220). See entry 266.

49 BARBER, Francis Morgan, 1845 . *Lecture on Drifting and*

Automatic Movable Torpedoes, Submarine Guns and Rockets. Newport, USN Torpedo Station, 1874. 40 pp., plates. Cover title *Lecture on Movable Torpedoes, 1875.*

Discusses torpedo use from the 15th century to 1873.

50 ________. *Lecture on Submarine Boats, and Their Application to Torpedo Operations.* Newport, U.S. Torpedo Station, 1875. 39 pp., plates. LC 15-9990 Revised.

References to David Bushnell, and notes on diving bells are included in the discussion.

Barincou, Robert. See entry 993.

Barjot, Admiral. See entry 1065.

51 BARNES, John Sanford, 1836-1911. *Submarine Warfare, Offensive and Defensive, Including a Discussion of the Offensive Torpedo System, Its Effects Upon Iron-Clad Ship Systems, and Influence Upon Future Naval Wars.* N.Y., Van Nostrand, 1869. 233 pp., including appendices. Frontis., diagrams, plates. LC 55-50504.

Preface is datelined "Naval Academy, June, 1868" and was thus finished three years after the Civil War in which torpedoes, mines and submarines had been employed successfully, both offensively and defensively. First attempt at a comprehensive compendium of information and reviews the past and extant information on the topics mentioned. This is an important landmark work in submarine history. A state-of-the-art survey. Barnes was a Lieutenant Commander at the time he compiled the book.

52 BARNES, Robert Hatfield. *United States Submarines.* 2nd. ed., New Haven, Conn. Morse, 1945. 221 pp., illus., ports, facisms, diagrs. LC 45-6060.

Reprinted mostly from articles which had been published in the *Connecticut Circle* magazine. Barnes was an officer in the USN and served for some time as public information officer at the USN Submarine Base in New London. The writing reflects the public relations background, however it is interesting and useful.

53 BARROWS, Nathaniel A., 1898?-1949. *Blow All Ballast! The Story of the Squalus.* N.Y., Dodd, Mead, 1940. xiv, 298 pp., frontis., illus., plates, ports. LC 40-27428.

Also published by the Court Book Co. in 1941.

Nat Barrows was a newspaper reporter for the Boston *Globe* who was on the scene during the salvage operations to raise the *Squalus.* His obituary written by R. Trumbull appeared in the N.Y. *Times* for July 13, 1949. There is a portrait and illus. in the obit.

54 BARTON, Otis. *The World Beneath the Sea.* N.Y., Crowell, 1953. 246 pp., illus. LC 52-13123.

Deals with submarine diving and the fauna of the sea.

Bartram, Robert. See entry 560.

55 BARTSCH, Max. *Was Jeder vom Deutschen U-Boot Wissen Muss.* Berlin, Limpert, 1940. 40 pp., illus. LC AF 48-3598*

56 BARTZ, Karl, 1900- . *Zeebrügge. Der Englische Angriff auf die Deutsche U-Boot-Basis.* Berlin, Im Deutschen Verlag, 1938. 193 pp., illus., diagrams, ports. LC 38-31482.

Concerns the Zeebrugge-Ostend Raids of 1918, and the British cruiser HMS *Vindictive.*

Baschieri, Francesco. See entries 1056, 1957.

Bascom, Willard. See entry 497.

La Base Navale du Havre et la Guerre Sous-Marine Secrète en Manche, 1914-1918. See entry 210.

57 *The Base Newspaper.* U.S. Submarine Base, New London, Connecticut. 1918-1921. (Newspaper) LC 47-33081.

First issue was July 1, 1918. Suspended publication in Dec. 1921. Title varied and during the period it was also issued as: *The Submarine Base Journal,* and *The Submarine Base Flotilla Journal.* Deals with news of training activities, sports and social events.

Basic Enlisted Submarine Text. (NavPers 10490). See entry 1395

Basic Scuba. See entry 1039.

58 BASS, George Fletcher. *Archaeology Under Water.* N.Y., Praeger, 1966. 224 pp., illus., maps. LC 66-12992 [66-73022].

Also published in London by Thames Publishing Co.

59 BASS, George Fletcher and others. *Cape Gelidonya: A Bronze Age Shipwreck.* Phila., American Philosophical Society, 1967. 177 pp., illus., maps. LC 67-28643.

Scholarly report on an underwater archaeological investigation.

60 BASS, George Fletcher, ed. *A History of Seafaring Based on Underwater Archaeology.* N.Y., Walker & co., 1972. 320 pp., illus. with b&w and color photographs, plus 115 drawings and 22 maps. Bibliography.

61 BATE, Norman, 1916- . *What a Wonderful Machine is the Submarine: A Picture Story*. N.Y., Scribners, 1961. unpaged, illus. LC 61-13609.

A picture book for children between ages of 4 and 8.

Les Bateaux Hémi-Plongeurs. See entry 1256.

Les Bateaux Sous-Marins. See entry 414.

Bateaux Sous-Marins a Grande Vitesse Sous L'Eau. See entry 1001.

Les Bateaux Sous-Marins et Les Submersibles. See entry 380.

62 BATEMAN, Charles T. *U-Boat Devilry, Illustrating the Heroism and Endurance of Merchant Seamen*. London, Hodder & Stoughton, 1918. xvi, 175 pp. LC 19-6493.

Bates, L. M. See entry 538.

63 BATES, Lindell Theodore, 1890- . *Manoeuvering as a Defense Against Attack by Submarines: The Automatic Course Indicator*. N.Y., Mail and Express Job Press, 1918. 35 pp., illus., diagrams. LC 18-2994.

64 ________. *Sinuous Courses and the Automatic Course Indicator for Naval Warfare*. NY, The Submarine Defense Association, 1918. 63 pp., illus., diagrams, tables. LC 19-1421.

Bathe, Basil W. See entry 818.

Bathyfolages. See entry 876.

Bathyscaphe. See entries 123, 622, 623, 631, 825, 969, 970, 971, 1146, 1152, 1224, 1265.

Il Battello Sottomarino. See entry 491.

Battle Below. See entry 199.

The Battle for Inner Space. See entry 1170.

The Battle of the Atlantic. See entry 517.

Battle Stations! See entry 637.

Battle Submerged. See entry 253.

Der Bau von Unterseebooten auf der Germaniawerft. See entry 1221.

65 BAUER, Hermann, 1875- . *Als Führer der U-Boote im Weltkriege: der Eintritt der U-Boot-Waffe in die Seekriegführung.* Leipzig, Koehler & Amelang, 1942. 470 pp., diagrams, map. LC AF 48-195*.

66 ________. *The Submarine, Its Importance as Part of a Fleet. Its Position in International Law, Its Employment in War, Its Future.* Translated by Hyman G. Rickover. Newport, R.I., USN War College, 1936. Translation of Entry 67.

67 ________. *Das Unterseeboot: Seine Bedeutung als Teil Einer Flotte, seine Stellung im Völkerrecht, seine Kriegsverwendung, seine Zukunft.* Berlin, E. S. Mittler & sohn, 1931. 140 pp., frontis. portrait of Admiral Bauer. LC 31-33400.

68 ________. *Reichsleitung und U-Bootseinsatz 1914 bis 1918: Zusammenarbeit Zwischen Politischer und Militärischer Führung im Kriege.* Lippoldsberg, Klosterhaus-Verlag, 1956. 111 pp. LC 57-25095‡.

Bauer, Wilhelm Sebastian Valentin, 1822-1876. See entries 106, 493, 582, 1098, 1229, 1230, 1419.

69 BAXTER, James Phinney, 3rd, 1893-1975. *Scientists Against Time.* Boston, Little, Brown, 1946. (Atlantic Monthly Press book series: Science in World War Two) xv, 473 pp., illus., maps. LC 46-7204.

Translated into French as *Secrets de la Science Americaine.* (Paris, 1947).

70 BAXTER, Richard, 1880- . *Stand By To Surface.* London, Cassell, 1944. 208 pp., plates. LC 44-7119.

Bayly, Sir Lewis. See entries 182-184.

71 BEACH, Edward Latimer, 1918- . *Around the World Submerged: The Voyage of the Triton.* NY, Holt, Rinehart & Winston, 1962: London, Hodder, 1964. 293 pp., illus. LC 62-18406.

The USS Triton (SSN 586) was commanded by Capt. Beach on the submerged cruise which followed Magellan's track. The cruise ended May 10, 1960.

72 ________. *Dust on the Sea.* NY, Holt, Rinehart & Winston, 1972. 351 pp. LC 79-15503. ISBN 0-03-076390-8.

A novel re submarine pack in the Pacific during World War Two.

73 BEACH, Edward Latimer, 1918- . *Run Silent, Run Deep.* NY, Holt, 1955. 364 pp. LC 55-6416.

Author served as USN submarine officer in Pacific during WW II, was later Naval Aide to President Eisenhower, and was commanding officer of the USS *Triton* during historic submerged world cruise in 1960. This book was made into a successful film with the same title. The movie starred Clark Gable and Burt Lancaster and was released by United Artists in 1958.

74 ———. *Submarine!* NY, Holt, 1952. 301 pp. LC 52-6623.

Personal account of WW II submarine warfare in the Pacific.

75 ———. *USS Triton, SSRN 586, First Submerged Navigation, 1960.* Washington, U.S. Government Printing Office, 1960. (2), 80 pp., illus. LC 60-61272.

The official log of the cruise.

Beach, Edward Latimer (Ned), 1918- . See entries 76, 304, 305, 578, 1419.

Biographical references may be found in the following works:

Becker, Beril—*Captain Edward L. Beach* Chicago, Encyclopaedia Britannica, 1963.

Current Biography Annual volume for 1960.

Christian Science Monitor May 24, 1960. p. 6, port.

N.Y. Herald Tribune August 19, 1958. p. 10, port.

NY Times August 20, 1958. p. 3, port.

NY Times—*Men in the News-1958.* Lippincott, 1959. p. 34.

Wagner, Frank—*Famous Underwater Adventurers.* NY, Dodd, 1962. pp. 125-134. Port, bibliog.

Who's Who in America. Chicago, Marquis, 1960/61 and later.

Wood, A. "New Creative Writers" in *Library Journal* February 1, 1955. 80:289, port.

The Beachhead Spies. See entry 1193.

Beall, M. E. See entry 746.

Beating the U-Boats (1917-18). See entry 214.

Beck, Pvt. Grafton D., 1852- (printer). See entry 3, 360.

76 BECKER, Beril. *Captain Edward L. Beach: Around the*

World Underwater. Illus. by Richard Mlodock. (Britannica Bookshelf: Great Lives for Young Americans) Chicago, Encyclopedia Britannica, 1963. 190 pp.

77 BECKER, Willi, 1893- . *Der Lusitania-fall und die deutsche Wissenschaft*. Greifswald, J. Abel, 1919. 86 pp. LC 24-12193.

78 BECKH, Albrecht. *Tauchkunst der Moderne Sport*. Einbandentwurf und Zeichnungen von Lilo und Walter Ganshorn. Munich, Braun & Schneider, 1952. 198 pp., illus. LC 53-17332.

79 BECKMANN, A. *Les Sous-Marine Allemands Devant New York*. (Mémoires, études et Documents Pour Servir a L'Histoire de la Guerre Mondiale) Paris, Payot, 1931. 175 pp.

80 ________. *U-Boote vor New-York: Die Kriegsfahrt eines Duetschen Unterseebootes nach Amerika*. Stuttgart, Franckh, 1931. 54 pp., illus., maps. LC 32-8439.

Bedford, U.S.S. See entry 1012.

The Bedford Incident. See entry 1012.

81 BEEBE, Charles William, 1877-1962. *Beneath Tropic Seas: A Record of Diving Among the Coral Reefs of Haiti*. N.Y., London, Putnam, 1928. xiii, 234 pp., illus., plates, ports, map. LC 28-22400.

"A few highlights of the tenth expedition of the Department of Tropical Research of the New York Zoological Society, making a study of the coral reefs and the fishes of the bay of Port-au-Prince, Haiti."

82 ________. *Half Mile Down*. (New ed.) NY, Duell, Sloan and Pearce, 1951. xxiii, 344 pp., illus., ports, map. LC 51-10409.

Diving operations at Nonsuch, Bermuda, sponsored by N.Y. Zoological Society. "History of Diving Bibliography" by Jocelyn Crane on p. 251.

83 ________. *A Media Milla de Profundidad*. Madrid, Espasa-Calpe, 1939. 282 pp., illus., plates, ports, diagrams. LC 44-31812.

84 ________. *Neunhundertdreiund zwanzig Meter Unter dem Meeresspiegel*. Translated by Max Müller. Leipzig, Brockhaus, 1935. 255 pp., plates, ports, diagrams. LC 50-45306.

Beebe, Charles William, 1877-1962. See entry 1419.

The Beginning of Modern Submarine Warfare. . . . See entries 2 and 3.

Die Bekanntmachungen des U-Boot-Handelskrieges. See entry 636.

Bekker, Cajus D. *pseud.* of Hans Dieter Berenbrok, 1924- .

85 BEKKER, Cajus D., *pseud. Augen Durch Nacht und Nebel.* 2nd ed. Hamburg, Stalling, 1964. 284 pp., illus., diagram.
Story of radar. Title translates to *Eyes Through Night and Fog.*

86 ________. *Defeat at Sea: The Struggle and Eventual Destruction of the German Navy, 1939-1945.* NY, Holt, 1955. 222 pp., illus. LC 55-9875.
Published in England with title *Swastika at Sea.*

87 ________. *Einzelkaempfer auf See.* Hamburg. Stalling, 1968. 210 pp., illus., diagrams.
German midget submarines and frogmen.

88 ________. *K-Men: The Story of the German Frogmen and Midget Submarines.* London, Kimber, 1955. 202 pp., plates.

89 ________. *K-Men: The Story of the German Frogmen and Midget Submarines.* London, G. Mann, 1973. 202 pp., illus. (Reprint of the 1955 edition). ISBN 0-7041-0006-1. LC 74-162011.

90 ________. *Kampf und Untergang der Kriegsmarine.* Hannover, Sponholz, 1953. 279 pp., illus.
The original work from which the translation for entry 86 was made.

91 BELKNAP, Reginald Rowan, 1871- . *The Yankee Mining Squadron: or, Laying the North Sea Mine Barrage.* Annapolis, U.S. Naval Institute, 1920. 110 pp., illus., map, diagrams. LC 20-13137.

Belknap, Reginald Rowan, 1871- . See entry 1053.

Below the Surface. See entry 215.

Below the Surface. See entry 589.

Belsheim, R. O. See entry 1389.

Ben Franklin (research submersible). See entries 972, 1442.

Bendick, Jeanne. See entry 991.

Beneath the Seven Seas. See entry 803.

Beneath Tropic Seas. See entry 81.

92 BENHAM, Clarence. *Diver's Luck: A Story of Pearling Days.* Sydney, Angus and Robertson, 1949. xiii, 258 pp., plates. LC 49-27682*.

93 ________. *Pearl Diver's Luck.* (1st American ed.) NY, Norton, 1950. 286 pp. LC 50-6063.

94 BENNETT, Mark H. J. *Under the Periscope.* London, Collins, 1919. 254 pp. LC 19-12821.
Prefatory note by L. Cope Cornford.

95 BENNETT, Peter Brian and David Hallen ELLIOTT, eds. *The Physiology and Medicine of Diving and Compressed Air Work.* London, Bailliere, 1969. Baltimore, Williams and Wilkins, 1969. 532 pp., plates, illus.

Benson, James. See entries 1425-1429.

96 BENTLEY, O. John. *The Thresher Disaster. The Most Tragic Dive in Submarine History.* NY, Doubleday, 1974. 372 pp., illus., appendices. ISBN 0-385-09909-6 LC 73-20507.
Actual publication date was January 10, 1975.

Berenbrok, Hans Dieter, 1924 See Bekker, Cajus D.

97 BERG, Ernst, 1886- . *Die Seeminen im Krieg, nach den Verhandlungen und Beschlüssen der Zweiten Haager Friedenskonferenz.* Frankenthal, Gohring, 1910. 81 pp. LC 43-37544.

98 BERGE, Victor. *Le Danger C'est ma Vie.* Paris, Editions de Paris, 1952. Translated from the Swedish by Marguerite Gay et Gerd de Mautort. 251 pp., illus. LC 53-37913.
Pearl diving.

99 ________. *Faran är Mitt Liv.* Stockholm, Folket i Bilds Förlag, 1953. 285 pp., illus. LC 54-36133.

100 ________. *Pearl Diver.* Rewritten by Henry Wysham Lanier from a transcript of Berge's words. Decorations by

Stephen Haweis. Jacket and frontispiece by Nicholas F. Riley. NY, Doubleday, Doran, 1931. 352 pp., illus. LC 31-29169.

101 ________. *Der Perlentaucher.* Frankfurt a M., Rutten & Loening, 1932. 317 pp. LC 32-14128.
Translated from the English edition by A. Normann.

Bergen, Claus, 1885- (Marine artist and painter of German submarines at sea and in port. One of his paintings hangs in a building at the USN Submarine Base in New London, Connecticut.) See entries 599, 902, 904, 905, 1072.

Bergland, Eric. See entry 270.

102 BERGMAN, P. G., and others, eds. *Physics of Sound in the Sea.* NY, Gordon and Breach, 1968. 4 parts in 3 volumes.
Part I. "Transmission" ed. by P. G. Bergman and A. Yaspan.
Part II. "Reverberation" ed. by E. Gerjuoy and A. Yaspan.
Part III. "Reflections of Sound from Submarines and Surface Vessels" ed. by J. K. Major.
Part IV. "Acoustic Properties of Water" ed. by R. Wildt.

Bernard, J. F. See entry 923.

103 BERNOTTI, R. *La Guerra Sui Mari. Nel Conflitto Mondiale, 1941-43.* Livorno, 1949. Maps, illus.

104 BERRES, Frances & others. *Submarine Rescue.* San Francisco, Wagner, 1959. 73 pp. LC 59-41988.
A book for the juvenile reader.

Berry, Erik, *pseud.* See entry 105a.

Bertin, Léon. See entry 333.

105 BERTRAND, Simon. *Le Sous-Marine: Sa Description, son Fonctionnement.* Paris, Imprimerie Busson, 1932. 16 leaves, illus., diagrams.

105a BEST, Ailena Champlin (Mrs. Herbert Best). *Underwater Warriors: Story of the American Frogmen.* By Erik Berry, pseud. NY, McKay, 1967. 152 pp., illus. LC 67-18004.
For the juvenile reader.

Beswick, John. See entry 670.

106 BETHGE, Hans-Georg. *Der Brandtaucher.* Bielefeld and Berlin, Delius, Klasing & co., 1968. 173 pp., illus., diags.

Detailed story of Wilhelm Bauer and the *Brandtaucher*, Germany's first submarine which he designed.

Bewaffnete Handelsschiffe im Weltkriege. See entry 1104.

Bezwungene Tiefen. See entry 710.

Bibliographic Sourcebook of Compressed Air, Diving and Submarine Medicine. See entry 616.

Bibliography. See entries 16, 122, 123, 157, 616, 628, 659, 660, 666, 694, 892, 894, 907, 948, 1066, 1081, 1197, 1277, 1324, 1326, 1491.

Bibliography of Submarine Warfare. See entry 659.

Bibliography of the Literature of Submarines. See entry 1081.

A Bibliography of Underwater Archaeology. See entry 694.

The Big Book of Real Submarines. See entry 809.

107 BILLINGS, Henry, 1901- . *Man Under Water.* NY, Viking Press, 1954. 189 pp., illus. LC 54-9597.

Deals with submarines and diving. Illustrated by the author.

108 BIRD, Peter. *Skin Diving.* (A Sportshelf Book) London, Soccer, 1963. 92 pp., illus.

Birmingham, Lloyd. See entry 1442.

109 BIRNBAUM, Karl E., 1924- . *Peace Moves and U-Boat Warfare: A Study of Imperial Germany's Policy Toward the U.S. April 18, 1916-January 9, 1917.* Hamden, Archon Books, 1970. xxi, 388 pp. SBN 208-00908-6. LC 78-95021.

A reprint of the 1958 edition

The Birth and Development of the American Submarine. See entry 175.

Bis Zum Letzten Flaggenschuss. See entry 1261.

110 BISCHOFF, Ernst, 1868- . *Der Morderversuch an Oberleutnant z.s. Crompton.* Umschlagzeichnung von Olaf Gulbransson. Leipzig, Hirzel, 1917. 31 pp., plates. LC 30-6918.

111 BISHOP, Farnham, 1886-1930. *The Story of the Submarine.* 1st ed. NY, The Century co., 1916. 211 pp., illus.

112 ________. *The Story of the Submarine.* Revised and Enlarged edition. NY, The Century co., 1929. xv, 228 pp., illus., ports. LC 29-11298.

Also reprinted in 1943.

113 BISHOP, Reginald. *Ships and Sailors of the Red Navy.* London, 1944.

114 BLACK, Adam Elliot. *Drill Book for the Use of Sub-marine Mining Engineer Volunteers.* London, Clowes, 1887. 91 pp., 54 plates. LC 18-27436.

Black Saturday. See entry 819.

115 BLACKBURN, J. A. and K. WATKINS. *The British Submarine in Being.* London, Gieves, 1920(?) 191 pp., illus.

116 BLACKMAN, D. J., ed. *Marine Archaeology.* (Colston Research Paper No. 23) Hamden, Archon Books, 1973. 532 pp., illus., maps. ISBN 0-208-01345-8.

117 BLAIR, Clay, Jr., 1925- . *The Atomic Submarine and Admiral Rickover.* NY, Holt, 1954. 277 pp., illus. LC 53-9584.

Published before the launching of the USS *Nautilus* (SSN 571) the first atomic submarine.

118 ________. *The Atomic Submarine.* London, Oldhams Press, 1955. 224 pp., illus. Frontis. port of Admiral Rickover.

118a ________. *Silent Victory: The United States Submarine War Against Japan.* Philadelphia, Lippincott, 1975. 1072 pp., illus.

". . . There will never be a rival book."—Edward L. Beach.

Blair, Clay, Jr., 1925- . See entries 18, 21, 22, 23.

Blaison, Louis. See entry 544.

119 BLAKE, R. F. *Submarine Signaling—The Protection of Shipping by a Wall of Sound and Other Uses of the Submarine Telegraph Oscillator.* (in Smithsonian Institution. *Annual Report. 1915.* Washington, 1916. pp. 203-213. illus.)

Reprinted from the *Proceedings* of the American Institute of Electrical Engineers. October 1914. LC 16-16010.

120 BLANCHON, Georges. *Les Sous-Marins et la Guerre Actuelle.* Paris, Bloud et Gay, 1915. 38 pp., illus. LC 15-14860.

121 BLECHMAN, Barry B. *The Changing Soviet Navy.* Washington, The Brookings Institution, 1973. 51 pp.

Blockade. See entries 756, 804, 805, 944, 1113, 1165-1167, 1413.

Le Blocus de L'Allemagne. See entry 1413.

Le Blocus et las Guerre Sous-Marine (1914-1918). See entry 756.

Blow All Ballast! See entry 53.

Blow Negative. See entry 1182.

The Blue Continent. See entries 1004, 1005.

Bock, Gerd. See entry 1038.

Boehmer, Robert. See entry 825.

Boesen, Victor. See entry 695.

Boevoi put Sovetskogo Voenno-Morskogo Flota. See entry 974.

Bond, George F. See entry 1393.

122 BOND, Guyla. *Nuclear-Powered Submarines, 1950-1965. A Checklist of Periodical Articles.* n.p., Guyla Bond, 1966. 118 pp. typescript with subject index and 10 pp. supplement.

123 ________. *Voyage to the Bottom, 1955-1965. A Checklist of Periodical Articles on Bathyscaphes and Mini-Subs, Man-in-the-Sea Projects and Flying Submarines.* n.p., Guyla Bond, 1966. 51 pp., typescript. Subject index.

124 BONEHILL, Ralph. *The Young Naval Captain: or, The War of All Nations.* Chicago, Thompson and Thomas, 1902. 221 pp., illus.

Highly melodramatic juvenile fiction. Young USN officer builds and commands Holland-type submarine against numerous hostile navies, including Japan and Russia.

125 BONHAM, Frank. *War Beneath the Sea.* NY, Crowell, 1962. 263 pp., glossary, bibliog. LC 62-12811.

Teen-age fiction re WW II submarine operations.

126 BONNER, Emil. *Die Helden der U-Boote.* Halle a.S., Marhold, 1935. 16 pp., (Deutschland ist Erwacht! Schriftenreihe fur die Schuljugend, Heft 7) LC 48-34313*.

The Book of Submarines. See entry 810.

Borden, Raymond Davis, 1877- . See entry 160.

127 BORELLI, Giovanni Alfonso, 1608-1679. *De Motu Animalium.* Rome, Angeli Bernabo, 1680.

Contains a chapter "Navis Urinatoriae Fabrica et Usus Exponitur" which deals with the diving bell.

128 BORGESON, Lillian (Mrs. Griffith B.) and Jack SPEIRS. *Skin and Scuba Diver.* (Arco Sports Library) NY, Arco; Toronto, Ambassador, 1963. 143 pp., illus. LC 62-16792.

129 BORGHESE, Junio Valerio, 1906- . *Sea Devils.* London, Melrose, 1952. Chicago, Regnery, 1954. 262 pp. Appendix. LC 54-4054.

Translated from the Italian *Decima Flottoglia Mas* by James Cleugh.

Borja, Robert. See entry 19.

130 BOSWORTH, Allan Rucker, 1901- . *Full Crash Dive.* NY, Duell, Sloan and Pearce, 1942. 27 pp. LC 42-5131.

First appeared in *Argosy* under title "Murder Goes to Sea."

131 ________. *My Love Affair With the Navy.* Foreword by Ernest M. Eller. NY, Norton, 1969. 288 pp. SBN 393-07449-8. LC 73-77397.

An informal, anecdotal memoir of USN history including much on submarines, and experiments in ocean bottom living.

132 BOURGOIS, Siméon, 1815-1887. *Mémoire sur la Navigation Sous-Marine Appliquée à la Défense des Ports.* Paris, 188-. 83 pp., folding plan. LC CA 17-378 Unrev'd.

133 ________. *Le Plongeur de Bourgois et Brun.* Paris?, 1877.

Bourgois's collaborating engineer and designer was Charles Brun, 1821-1888.

134 BOURNE, William, d. 1583. *Inventions and Devices.* London, 1578.

135 BOWER, John Graham, 1886- . *The Story of Our Submarines,* by Klaxon, *pseud.* Edinburgh and London, Blackwood, 1919. 297 pp. LC 20-494 Revised.

The Boy Allies With the Navy series. See entry 587.

Boy Scouts in the North Sea. See entry 1010.

The Boys' Book of Submarines. See entry 240.

136 BOYCOTT, George Walter Morgan. *Compressed Air Work and Diving: A Handbook for Engineers, Comprising Deep Water Diving and the Use of Compressed Air for Sinking Caissons and Cylinders and for Driving Subaqueous Tunnels.* London, Lockwood and Son, 1909. xii, 116 pp., illus. tables, folding plates. LC 9-30881.

137 BRADFORD, Royal Bird, 1844-1914. *History of Torpedo Warfare.* Newport, USN Torpedo Station, 1882. 90 pp.

Use of torpedoes from time of David Bushnell through the South American war of 1879-80.

Bradley, Humphrey. See entry 569.

138 BRADY, Edward M. *Marine Salvage Operations.* Cambridge, Md., Cornell Maritime Press, 1960. 237 pp., diagrams. LC 59-12836.

A practical treatise on diving and underwater work.

139 BRAGADIN, Marc Antonio. *The Italian Navy in World War Two.* Annapolis, Naval Institute Press, 1957. 380 pp., illus, appendix.

Der Brandtaucher. See entries 106 and 1098.

140 BRASE, W. A. H. *Unterseeboote.* (Herausgegeben im Auftrage der Submarine Gesellschaft.) Hannover, Duisberg 1933. 26 leaves.
(Submarine Bibliographie. Fachgruppe 19, Folge 1)

141 BRASSEUR, Pierre. *La Guerre de Mines.* Paris, Nizet et Bastard, 1939. 155 pp., bibliography.

A thesis submitted to University of Paris.

142 BRAVETTA, Ettore, 1862- . *Sottomarini, Sommergibili e Torpedini.* Milano, Fratelli Treves, 1915. 230 pp., illus.,

diagrams.

Bravetta, Ettore. See entry 702.

Brazil. See entries 1294, 1301, 1312, 1318.

143 BREHMER, Artur. *Die Kuhne Fahrt der Deutschland. Im Handels-U-Boot von Bremem bis Baltimore und Zuruck.* Berlin, B. Siegmund, 1916. 96 pp.

Voyage of the "merchant submarine" *Deutschland.*

144 BRENNECKE, Hans Joachim. *The Hunters and the Hunted.* Translated by R. H. Stevens. NY, Norton, 1958. London, Burke, 1958. Toronto, Ambassador, 1958. 320 pp., illus. LC 58-13949.

145 BREYER, Siegfried. *Guide to the Soviet Navy.* Translated by LCDR. M. W. Henley, R.N. (ret) Annapolis, U.S. Naval Institute, 1970. 353 pp., illus., maps. ISBN 0-87021-237-0. LC 70-97170.

Bridge, Sir Cyprian Arthur George, 1839-1924. See entry 328.

146 BRIGGS, Peter. *Men in the Sea.* NY, Simon & Schuster, 1968. 128 pp., illus. LC 68-18967.

A book about oceanography and oceanographers for young folks age twelve and up. Research submarine *Alvin.* Biographical sketches of: Athelstan Spilhaus, Roger Revelle, Maurice Ewing, Edwin Link, George Bond, Anton Bruun, Townsend Cromwell, Columbus O'Donnell Iselin, and Tap Pryor.

147 BRIGGS, Raymonde *New Submersible Battle Cruiser with Sixteen Inch Guns.* NY, Briggs, 1940. 28 pp., illus. LC 42-16178.

Describes the vessel, work of the inventor Nevil Monroe Hopkins, 1873-.

148 ________. *The Super Submarine, the Only Vessel Now Capable of Safely Carrying and Servicing a Sixteen Inch Gun.* Bogota, N.J., The Dancey Printing Co., 1942. 29 pp., illus., plate.

149 BRIGHTON, Ray. *They Came to Fish.* Portsmouth, N.H., "Portsmouth 350," 1974. Two volumes.

History of Portsmouth including information on the Portsmouth Navy Yard and Portsmouth-built submarines.

Brightwell, L. R. See entry 842.

Britain's Clandestine Submarines, 1914-1915. See entry 1143.

British Merchant Marine. See entry 646.

British Navy. See entry 37.

150 British Sub-Aqua Club. *Underwater Diving Manual: Official Instructions on Underwater Swimming Techniques* by George F. Brookes and Alan V. Broadhurst. 3rd revised edition. London, S. Paul, 1962. 240 pp., illus.

British Sub-Aqua Club. See entry 1482.

The British Submarine. See entry 784.

The British Submarine In Being. See entry 115.

British Submarines. See entry 772.

British Submarines. See entries 7, 9, 25, 41, 70, 94, 115, 155, 156, 190-194, 200, 342, 358, 359, 384, 483, 507, 510, 551-554, 571, 572, 583, 584, 637-642, 673, 679, 680-684, 686, 689, 699, 700-704, 814, 832, 938, 1047, 1115, 1143, 1231, 1272, 1425-1429, 1489, 1490.

British Submarines at War. See entry 832.

British Warships, 1914-1919. See entry 315.

Broadhurst, Alan V. See entry 150.

Brochett, Eleanor. See entry 1056.

151 BRODIE, Bernard, 1910- . *Sea Power in the Machine Age.* 2nd ed. Princeton University Press, 1943. London, Oxford University Press, 1944. Toronto, Saunders, 1944. 462 pp. LC A43-2763.

Note: the Oxford University Press edition has only 255 pp.

152 BRONSON-HOWARD, George, 1921- . *Handbook for Skin Divers.* Greenwich, Conn., Fawcett Publications, 1956. 144 pp., illus. LC 56-3792.

Brookes, George F. See entries 150, 849.

Broome, J. E. See entries 183, 184.

Broomfield, Peter. See entry 494.

153 BROU, Willy Charles, 1914- . *Combat Beneath the Sea.* Translated from the French by Edward Fitzgerald. NY, Crowell, 1957. 240 pp. LC 57-10109.

Midget submarines of World War Two.

154 ________ *War Beneath the Sea.* London, Muller, 1958. 239 pp., plates.

Also published in Toronto by Saunders.
Same book as entry 153 with variant title.

Brown, Alec. See entries 499, 500.

Browne, Gordan. See entry 239.

Bruce, Gordon. See entry 17.

Brun, Charles, 1821-1888. See entry 133.

Brundza, Paul. See entry 1172.

Brunner, Karl Eugen. See entry 268.

155 BRYANT, Benjamin, 1905- . *One Man Band: The Memoirs of a Submarine Commanding Officer.* London, Kimber, 1958. 238 pp., plates. LC 59-20714.

World War Two submarine operations. British submarines *Safari* and *Sealion.*

156 ________. *Submarine Commander.* NY, Ballantine Books, 1960. 224 pp.

Paperback edition of *One Man Band.*

Bryant, Benjamin, 1905- . See also entry 365.

157 BRYCE, Barbara Ann. *An Annotated Literature Survey of Submarines, Torpedoes, Anti-Submarine Warfare, Undersea Weapon Systems, and Oceanography: 1941 to January 1962.* Los Angeles, Autonetics, A Division of North American Aviation, Inc. (Project EM-11-62-102), 9 March 1962.

Unpaged, processed. 485 references. Indexed three ways: author, source and corporate author, and subject. Unindexed appendix of 48 references. Unclassified, limited distribution. Also for sale by Office of Technical Services as AD 273660.

158 BUCHARD, Henri Pierre Gabriel, 1884- . *Torpilles et Torpillieurs des Nations Étrangères.* (Bibliotheque du Marin) Paris, Berger-Levrault, 1889. 222 pp., plates. LC 6-25004.

158a BUCHEIM, Lothar-Günther. *The Boat.* NY, Knopf, 1975. 463 pp. ISBN 0-394-49105-X LC 74-21314

A novel about a World War Two U-Boat and her crew.
Translated by Denver & Helen Lindley. End paper illus. shows cutaway of the U-boat.

159 BUHLER, Ottmar. 1884- . *Neutralitat, Blockade/ und U-Bootkrieg in der Entwicklung des Modernen Volkerrechts.* Berlin, Junker und Dunnhaupt, 1940.

Bullhead, USS (SS 332). See entries 1126, 1127.

Bullock, Michael. See entry 631.

Los Buques Submarinos Actuales. . . . See entry 787.

160 BURANELLI, Prosper, 1890- . *Maggie, of the Suicide Fleet.* As written from the log of Raymond D. Borden, Lieutenant, USNR, by Prosper Buranelli. Illustrated by Herb Roth. NY, Doubleday, Doran, 1930. 278 pp. illus. LC 30-9295.

Tales of a World War One submarine chaser.

161 BURKE, Edmund. *The Diver's World: An Introduction.* NY, Van Nostrand, 196? 224 pp., illus. Glossary.

Skin and SCUBA diving as a sport.

162 ________. *The Underwater Handbook.* London, Muller, 1963. 260 pp., plates.

163 BURGOYNE, Alan Hughes, 1880- . *Submarine Navigation Past and Present.* London, Richards; NY, Dutton, 1903. 2 volumes, illus., plates, bibliography. LC 3-23845.

An important source book by a submarine engineer.

Burton, Douglas. See entry 993.

Burton, H. M. See entry 313.

164 BUSCH, Fritz Otto, 1890- . *U-Bootsfahrten. Buchschmuck von Werner Chomton.* Berlin, Schneider, 1938. 95 pp., illus., maps. LC A50-1649.

165 ________. *U-Boots-Taten,* von korvettenkapitan a.d. Fritz Otto Busch (Peter Cornelissen) mit 16 kupfertiefdruckbildern, 6 gefechtsskizzen, 3 U-boots-schnittzeichnungen und 1 u-boots-tabelle. Berlin, Hobbing, 1934. 125 pp. illus. LC 35-6582.

166 BUSCH, Harald. *Meutes Sous-Marines.* Paris, France-Empire, 1953.

167 ________. *So War der U-Boots-Krieg.* Bielefeld, Deutscher Heimat, 1957. 472 pp., illus.

168 ________. *U-Boats at War.* Translated from the German by L. P. R. Wilson. London, Putnam, 1955. Toronto, McClelland, 1955. 286 pp., illus., appendix.

169 ________. *U-Boats at War. German Submarines in Action 1939-1945.* Translated by L. P. R. Wilson. NY, Ballantine Books, 1955. xii, 176 pp., appendices.

170 BUSH ROMERO, Pablo, 1905- . *Under the Waters of Mexico.* NY, Carlton, 1964. 184 pp., illus. Photos by Genaro Hurtado. Maps by Enrique Enriquez. LC 66-31654.

Underwater archaeology.

171 BUSHELL, T. A. *Eight Bells. Royal Mail Lines War Story, 1939-45.* London, 1945. 207 pp.

Bushnell, David, 1742?-1824? See entries 2, 3, 50, 51, 1241, 1255, and 1420.

172 BUSHNELL, Eddie, 1907- . *Underseas Log.* From the records of Eddie Bushnell, deep-sea diver, as told to M. O'Moran. Photography by William L. Morgan. Underseas photography by Bushnell. Caldwell, Idaho, The Caxton Printers, 1947. 266 pp., plates. LC 47-1927.

Deep sea diving.

173 BUSS, J. P. *Amerikanische Menschlichkeit im Lichte des Diplomatischen Notenwechsels.* Berlin, Schwetschke, 1916. 92 pp. LC 24-9941.

Busuttill, Mike. See entry 993.

A Buvarnaszad Haboru. See entry 692.

By Guess and By God. See entries 190, 191.

174 BYKHOVSKY, Izrail Adolfovich. *Atomnye Podvodnye Lodki.* (Atomic Submarines). Leningrad, 1957.

By Sea and By Stealth. See entry 1466.

Bywater, Hector C. See entry 372.

C

Cabirol's Diving Suit. See entry 354a.

175 CABLE, Frank Taylor, 1863-1945. *The Birth and Development of the American Submarine.* NY and London, Harper, 1924. 337 pp., plates, ports. Supplementary chapter (p. 314-337 by W. W. Kimball) LC 24-12751.

Important for information on the development of our embryo submarine fleet, and contributions of John P. Holland to submarine design. Cable was associated with Electric Boat company.

Cable, Frank Taylor, 1863-1945. Submarine expert. Biography with portrait and autograph appears in White's *National Cyclopaedia of American Biography* 1948. (Vol. 34, pp. 164-165).

Cageling, M. A. See entry 804.

176 CAIDIN, Martin, 1927- . *Hydrospace.* NY, Dutton, 1964. Toronto, Clarke, Irwin, 1964. 320 pp., illus., diagrams. LC 64-21862.

Submarines, ASW, oceanography. Popular survey of state-of-the-art.

177 ________. *The Last Fathom.* Des Moines, Meredith Press, 1967. 312 pp. LC 67-18496.

"An explosive novel of suspense beneath the sea."

178 CALLAHAN, Vincent Francis, 1931- , ed. *Underwater Defense Handbook.* 2nd ed. Washington, Callahan Publications, 1963. 138 pp., illus. LC 63-5594. Lois C. Philmus, contributing ed.

179 CALVERT, James Francis, 1920- . *Surface at the Pole: The Extraordinary Voyages of the USS Skate.* NY, McGraw-Hill, 1960. 220 pp., illus. LC 60-10597.

Written by the man who was Skipper of the USS Skate (SSN 578) the second USN submarine to operate under the polar ice.

Calvert, James Francis, 1920- biographical reference in *Men in the News-1958.* (Phila., Lippincott, 1959. p. 50) See also entry number 908.

180 CAMBRIDGE, Richard Owen, 1717-1802. *An Aerial Race Between a Briton and a German, Together with Brief References to a Submarine, From a Mock-Heroic Poem Printed in 1751.* London, Chiswick Press, 1918. 15 pp., frontis, illus. LC 19-12035.

These are extracts from Cambridge's *Scribleriad* (1751) which were edited by Howard Coppuck Levis and printed in an edition of 100 copies for private distribution.

181 CAMPAGNA, Enzo. *La Nave Subacquea: Sottomarini e Sommergibili.* Milan, Hoepli, 1915. 346 pp., illus., diagrams. LC 42-12470.

182 CAMPBELL, Gordon, 1886-1953. *Mes Navires Mystérieux.* Translated by André Cogniet. Preface by Vice Admiral Grasset and Admiral Sir Lewis Bayly. Paris, Payot, 1929. 260 pp., plates, ports. LC 30-1853.

183 ________. *My Mystery Ships.* Foreword by RADM William S. Sims. Intro. by ADM Sir Lewis Bayly. Illus. by J. E. Broome. NY, Doubleday, Doran, 1929. 318 pp., frontis., illus., plates, ports. LC 29-6355. "First Edition".

British "Q" ships and WW I anti-submarine operations.

184 ________. *My Mystery Ships.* Foreword by Sir Lewis Bayly and illus. by J. E. Broome. London, Hodder, 1936. 300 pp., illus. "Popular Edition."

Also reissued in their Black Jacket series with 310 pp. Also a paperback edition by Musson.

185 ________. *Wir Jagen Deutsche U-Boote.* Translated by E. Freiherr v. Spiegel from *My Mystery Ships.* Gutersloh, Bertelsmann, 1937. 234 pp., illus., ports. LC 51-46804.

Campbell, Gordon, 1886-1953. British Admiral, died in London October 5, 1953. Obituaries appeared in following publications:

Illustrated London News. port. 223:549. Oct. 10, 1953.
New York Times. port. p. 29 October 6, 1953.
Newsweek 42:77 October 19, 1953.
Time 62:104 October 19, 1953.
Wilson Library Bulletin. 28:328. December, 1953.

Canada. See entries 1143, 1313.

186 CANADA. Delegate to the Conference on the Limitation of Armament. Washington, D.C., 1921-1922. *Conference on the Limitation of Armament Held at Washington November 12, 1921, to February 6, 1922. Report of the Canadian Delegate . . .* Ottawa, F. A. Acland, printer, 1922. 222 pp. (Parliament 1922. Sessional paper, No. 47) LC 23-1798. Sir Robert Borden, delegate.

187 CANADA. Royal Commission on War Supplies Purchases. *Report of the Commissioner Concerning Purchase of Submarines.* Ottawa, Printed by J. de L. Tache, printer to the King, 1917. 25 pp. Sir Charles Davidson, Commissioner. LC 47-33082.

Canadian submarines. See entry 1271.

Canby, Courtlandt. See entry 473.

Cape Gelidonya: A Bronze Age Shipwreck. See entry 59.

The Captain and Submarine CSS Hunley. See entry 352.

Captain Cousteau's Underwater Treasury. See entry 263.

Captain Edward L. Beach: Around the World Underwater. See entry 76.

Carbon Dioxide Removal. See entry 1390.

Cargo Aircraft and Cargo Submarines. See entry 1305.

Carlisle, Madelyn. See entry 188.

188 CARLISLE, Norman V., 1910- and Madelyn CARLISLE. *About Submarines.* (About Books series) Illus. by Roger Herrington. Melmont Publications, 1969. 47 pp., illus. LC 69-14683.

Juvenile for grades 2 to 5.

189 CARMER, Carl Lamson, 1893- . *The Jesse James of the Java Sea.* NY, Rinehart. 1945. 119 pp. LC 46-87.

Exploits of USS Seawolf (SS 197) during WW II.

Carmichael, Sgt. Archibald, 1831- (printer). See entries 3, 360.

Carnaghan, Jean G. See entry 1345.

Carnes, Cecil. See entry 686.

Carney, R. B. See entry 817.

190 CARR, William Guy, 1895-1959. *By Guess and By God. The Story of the British Submarines in the War.* Preface by ADM S. S. Hall. London, Hutchinson, 1930. 288 pp., illus.

British submarine E-11.

191 ________. *By Guess and By God: The Story of the British Submarines in the War.* Preface by ADM S. S. Hall. NY, Doubleday, Doran, 1930. First American Edition. 310 pp., frontis, plates. LC 30-25100.

192 ________. *Good Hunting: Being Volume Three of By Guess and By God.* London & Melbourne, Hutchinson, 1940. 288 pp., plates, ports. LC 41-4662.

193 ________. *Hell's Angels of the Deep.* London, Hutchinson, 1932. 288 pp., frontis, illus., ports. LC 33-2647.

194 ________. *Out of the Mists: Great Deeds of the Navy in the Last War and Her Role Today.* London, NY, Hutchinson, 1942. 176 pp., frontis, plates. LC 42-25173.

Carr, William James Guy, 1895-1959. Commander in the Royal Canadian Navy, who served on active duty during both world wars. A writer on naval topics. Died in Toronto October 2, 1959. His obituary appears in:

New York Times October 6, 1959. p. 39

Wilson Library Bulletin December 1959. 34:247

195 CARRERO, Guillermo. *El Submarino Torpedero en el Ataque.* Madrid, Escelicer, 1941. 104 pp., tables, diagrams. LC 43-18244.

Carrier, Barbara. See entries 196, 197.

196 CARRIER, Rick. *Dive: The Complete Book of Skin Diving.* Barbara Carrier, joint author. NY, Funk, 1955. 204 pp., illus., map, bibliog. LC 55-5481 rev.

197 ________ and Barbara CARRIER. *Dive: The Complete Book of Skin Diving.* Revised by Gene Parker. NY, Funk, 1963, and Don Mills, Ontario, Longmans, 1963. 296 pp., illus. LC 63-9014.

Also published in London by Kaye in 1963 with title *The Complete Book of Skin Diving.* 298 pp., illus.

198 CARSON, Edward Henry Carson, *Baron*, 1854-1935. *The War on German Submarines. Sir Edward Carson on the British Navy's Success.* London, Unwin, 1917. 8 pp.

LC 17-14125 Revised.

Carter, K. E. See entry 1242.

Cartilla de Electricidad. . . . See entry 828.

Casabianca: The Secret Missions of a Famous Submarine. See entry 779.

Casabianca (French submarine). See entry 779.

Case, A. See entry 334.

Case, R. M. See entry 334.

199 CASEY, Robert Joseph, 1890-1962. *Battle Below: The War of the Submarines.* Indianapolis, Bobbs-Merrill, 1945. 380 pp., illus., glossary. LC 45-6451.

Chatty, informal history of a war correspondent. Not entirely historically accurate.

200 CASING, James. *Submariners.* London and Toronto, Macmillan, 1951. xxiv, 184 pp.

201 CASTEX, Raoul Victor Patrice, 1878- . *Synthèse de la Guerre Sous-Marine: de Ponchartrain à Tirpitz.* Paris, Challamel, 1920. 228 pp., diagrams. LC 22-2329.

202 ________. *Theories Strategiques.* Paris, 1929.

203 CAULFIELD, Max (Malachy Francis), 1915- . *Tomorrow Never Came: The Story of the SS Athenia.* NY, Norton, 1959. 223 pp., plates. LC 59-5755.

Originally published in England and Canada in 1958 with title: *Night of Terror: The Story of the Athenia Affair.*

204 CAYFORD, John E. *Underwater Work: A Manual of Scuba Commercial, Salvage and Construction Operations.* 2nd ed., rev. & enl. Cambridge, Md., Cornell Maritime Press, 1967. 272 pp., illus., bibliog. LC 66-28081.

For experienced divers by a man who has been engaged in underwater work for many years.

C Class Submarines. See entry 1377.

205 CENTRO DE ESTUDIOS DE DERECHO INTERNACIONAL PÚBLICO, Buenos Aires. *Proyecto de Codificación del Empleo del Submarino en la Guerra.* Buenos Aires, University of Buenos Aires, 1929. 63 pp. LC A30-483 Revised.

Ceylon. See entry 227.

CGR 3070. See entry 1238.

206 CHACK, Paul, 1876- . *Sur Mer . . . et Dessous.* Paris, Les Editions de France, 1938. 117 pp., illus. LC 40-2448.

The Challenge of the Sea. See entry 225.

Challenge of the Sea Floor. See entry 403.

Challengers of the Deep. See entry 281.

207 CHALMERS, William Scott, 1888- . *Max Horton and the Western Approaches.* London, Hodder & Stoughton, 1954: Toronto, Musson, 1954. 302 pp., plates, maps. LC 55-15430.

208 CHAMBLISS, William C. *The Silent Service.* (Signet 1658) NY, New American Library, 1959. illus.

Collection of tales re American submarines during WW II as shown on TV series "The Silent Service."

208 ________. *The Silent Service.* (Signet Book) NY, New American Library, 1974. Copyright 1959. 190 pp., illus. A re-issue.

The Changing Soviet Navy. See entry 121.

209 CHANUKOFF, Lon. *Submarine Z-1: A Novel of Life and Death at the Bottom of the Sea.* NY, Citadel Press, 1960. 221 pp. LC 60-13932.

First published in Yiddish around 1930. Later translated into Russian, and here for the first time in English. Antiwar polemic.

La Chasse aux Mines. See entry 511.

210 CHATELLE, Albert. *La Base Navale du Havre et la Guerre Sous-Marine Secrète en Manche, 1914-1918.* Paris, Editions Médicis, 1949. 261 pp., illus., ports, maps, LC 49-53387.

211 CHATHAM, ENGLAND. SCHOOL OF MILITARY ENGINEERING. *Rough Notes of Seven Lectures on Submarine Mining.* Chatham, School of Military Engineering, 1877. 32 pp., illus., diagrams. LC 17-9991.

212 ________. *Rough Notes of Twenty-One Elementary Lectures on Electricity as Applied to Submarine Mining. . . .* Chatham, School of Military Engineering, 1877. 39 pp., diagrams. LC 19-14288.

213 CHATTERTON, Edward Keble, 1878-1944. *Amazing Adventure: A Thrilling Naval Biography.* London, Hurst & Blackett, 1935: Toronto, Ryerson, 1935. 285 pp., illus., plates, port. LC 36-8586.

Narrating the naval career of CDR Godfrey Herbert.

214 ________. *Beating the U-Boats (1917-18).* London, Hurst & Blackett, 1943. 171 pp.

Sequel to his *Fighting the U-Boats.*

215 ________. *Below the Surface.* London, Hurst & Blackett, 1934 and 1935. 286 pp.

Based upon a play by J. L. F. Hunt and H. G. Stoker.

216 ________. *Danger Zone.* Boston, Little, Brown, 1934. 437 pp. LC 34-41068. London, Rich and Cowan, 1934. 437 pp. LC 34-41067.

217 ________. *Fighting the U-Boats.* London, Hurst & Blackett, 1942. Toronto, Ryerson, 1942. 216 pp., illus., maps, plans, plates. LC A42-2453.

Covers the period 1914-1916. See also entry 214.

218 ________. *Q-Ships and Their Story.* London, Sidgwick & Jackson, 1922. Boston, Lauriat, 1922. 276 pp., illus. LC 22-14727.

219 ________. *Q-Ships and their Story.* Annapolis, Naval Institute Press, 1972. 288 pp., illus.

Chatterton was an English journalist and writer who was born in Sheffield and educated at Oxford. He served in the navy from 1914-1918 during WW I.

220 CHENEY, Cora (Mrs. B. W. Partridge, Jr.) and Benjamin Waring PARTRIDGE. *Underseas! The Challenge of the Deep Frontier.* NY, Coward, McCann, 1961: Toronto, Longmans, 1961. 121 pp., illus. LC 61-8249.

Submarines, diving, and oceanography. Written for grades 5 to 9.

221 CHIEF OF NAVAL OPERATIONS. *ASW: Anti-Submarine Warfare.* Washington, CNO, 1959. 38 pp., illus.

Chimie Pneumatique. . . . See entry 424.

China, Republic of. See entry 1313.

Chomton, Werner. See entry 164.

222 CHRISTIERNSSON, Nils Henrik, 1883- . *Med Tyska*

U-Båtar och i Tyska Fångläger. Stockholm, Norstedt, 1917. 93 pp. LC 20-23581.

Ciampi, Elgin. See entry 1014.

The Cinderellas of the Fleet. See entry 931.

223 CLAFLIN, Wilbur A. *A Concise Project for Equalizing and Conquering the Submarine Menace, Submitted to the President and the People of the United States.* Montpelier? Vt., 1917. 10 numbered leaves. LC CA 18-225 Unrev'd.

The Clammer and the Submarine. See entry 626.

Clapper, G. P. See entry 952.

224 CLARK, William Bell, 1889- . *When the U-Boats Came to America.* Boston, Little, Brown, 1929. 359 pp., frontis, plates, bibliog. LC 29-20137.

"I have aimed to present an authentic history of the operations of German submarines in the western Atlantic during the last six months of the world war, together with a comprehensive account of their effect upon the United States."—Preface.

225 CLARKE, Arthur Charles, 1917- . *The Challenge of the Sea.* NY, Holt, Rinehart & Winston, 1960. Illus. by Alex Schomburg. Intro. by Wernher von Braun. 167 pp. (W)60-11244.

For the young reader.

226 ———. *The Deep Range.* (Signet Book) NY, New American Library, 1957. 175 pp.

"A thrilling novel of life under the sea one hundred years from now."

227 ———. *The Reefs of Taprobane: Underwater Adventures Around Ceylon.* With color and b&w photos by Mike Wilson. 1st ed. NY, Harper, 1957. 205 pp., illus. LC 57-6136.

Claus Bergen: Leben und Werk. . . . See entry 599.

Claustrophobic Reactions to Some Stresses of Submarine Service. See entry 1344.

228 CLERC-RAMPAL, Georges, 1870- . *La Marine Française Pendant la Grande Guerre (Août 1914-Novembre 1918).* Preface by M.-A. Lorin. Paris, Larousse, 1919. 221 pp., illus., plates, diagrams. LC 26-7731.

229 ________. *Les Sous-Marins.* Paris, Hachette, 1919. 184 pp., plates, diagrams. LC 20-7359.

Cleugh, James. See entries 129, 1257.

Coast Defense. See entry 1100.

Cobb, David. See entry 554.

230 COCCHIA, Aldo, 1900- . *The Hunters and the Hunted: Adventures of Italian Naval Forces.* Translated by M. Gwyer. Annapolis, U.S. Naval Institute, 1958. 179 pp., plates. LC 58-8018.

231 ________. *Submarines Attacking.* London, Kimber, 1956. Toronto, Ryerson. 1956. 204 pp., plates. LC 57-2134.

This is the same book as 230 with variant title.

Cocchia, Aldo. See also entry 987.

Cochino, USS (SS 345). See entry 762.

Codman, Anna Kneeland. See entry 417.

Cody, USS (fictional). See entry 361.

232 COGGINS, Jack. *Prepare to Dive! The Story of Man Undersea.* Illus. by the author. NY, Dodd, Mead, 1971. 128 pp., illus. chronology. ISBN 0-396-06384-5. LC 77-153893.

Coggins, Jack. See also entries 1037, 1184.

Cogniet, André. See entry 182.

233 COHEN, Paul. *The Realm of the Submarine.* NY, Macmillan, 1969. 274 pp. Endpapers are illustrated with submarine silhouettes. LC 69-12176.

Deals principally with non-military submarines. "The exciting story of modern underwater exploration and engineering and the role of today's submarines in the building of tomorrow's undersea empires."—from the jacket blurb.

234 COLBY, Carroll Burleigh, 1904- . *Frogmen: Training, Equipment, and Operations of Our Navy's Undersea Fighters.* NY, Coward-McCann, 1954. 48 pp., illus. LC 54-11357.

For the young reader.

235 ________. *Submarine: Men and Ships of the U.S. Submarine Fleet.* NY, Coward-McCann, 1953. 48 pp., illus. LC 53-12465. Reprinted in 1960.

For the young reader.

236 ________. *Underwater World: Exploration Under the Surface of the Sea.* NY, Coward-McCann, 1966. Toronto, Longmans, 1966. 48 pp., illus. LC 66-7615.

For the young reader.

237 COLDEN, Cadwallader D. *The Life of Robert Fulton.* NY, Kirk & Mercein, 1817. vi, 371 pp. Appendix, folding chart.

Concerned mainly with steamboats, but includes information on Fulton's work with submarines and torpedoes in France and England.

238 COLE, Robert Hugh, 1914- . *Underwater Explosions.* NY, Dover, 1965. 437 pp., illus. LC 65-22737.

Colegrave, E. H. M. See entry 578.

Colledge, J. J. See entry 315.

The Collier Quick and Easy Guide to Skin and Scuba Diving. See entry 1094.

239 COLLINGWOOD, Harry, *pseud. Log of the Flying Fish: A Story of Aerial and Submarine Peril and Adventure.* Illus. by Gordon Browne. London, Blackie, 1886. 384 pp., illus.

Fiction. Author's real name is William Joseph Cosens LANCASTER.

240 COLLINS, Archie Frederick, 1869- . *The Boys' Book of Submarines.* NY, Stokes, 1917. 220 pp., illus. LC 17-31264.

Virgil Dewey COLLINS, 1898- , joint author.

Columbus, Christopher. See entry 782.

Combat Beneath the Sea. See entries 153, 154.

Combat Training of Submarine Crews. See entry 1322.

241 COMITÉ VOOR DE VIERING VAN HET 40-JARIG BESTAAN HR. MS. ONDERZEEDIENST. *Veertig Jaren Onderzeedienst 1906-1946.* Amsterdam, Scheltens & Giltay, 1947. 208 pp., illus., ports. LC 54-20283.

Commemorating 40 years of the Dutch submarine service.

Commander Crabb. See entry 1002.

242 COMMITTEE ON UNDERSEA WARFARE. *Proceedings*

of the Symposium on Aspects of Deep Sea Research. Washington, National Academy of Science, 1957. ? pp., illus.

243 COMPERE, Tom, ed. *The Navy Bluebook.* NY, Military Publishing Institute, 1960. 374 pp., illus.

Navy Facts is bound with and consecutively paged. Much general information re the Navy and seapower. High percentage about submarines.

The Complete Book of Skin Diving. See entry 197.

The Complete Illustrated Guide to Snorkel and Deep Diving. See entry 767.

The Complete Manual of Free Diving. See entry 425.

Compressed Air Work and Diving. See entry 136.

A Concise Project for Equalizing and Conquering the Submarine Menace. See entry 223.

244 CONDROYER, Émile. *Les Pionniers de la Plongée: Histoire des Machines Plongeantes.* Paris, Peyronnet, 1948. 292 pp., bibliog. LC 50-23767.

The Confederate Privateers. See entry 1049.

Confederate submarines. See entries 51, 352, 854, 963, 1049, 1100, and 1149.

Conference on the Limitation of Armament . . . Washington . . . 1921-22. See entry 186.

245 CONFERENCE ON UNDERWATER ARCHAEOLOGY, ST. PAUL, 1963. *Diving Into the Past: Theories, Techniques, and Applications of Underwater Archaeology.* The Proceedings of a Conference Sponsored by the Minnesota Historical Society, St. Paul, April 26-27, 1963. Edited by June Drenning Holmquist and Ardis Hillman Wheeler. St. Paul, Minnesota Historical Society, 1964. 111 pp., illus. LC 64-63332.

Connett, Eugene V. See entry 1039.

246 CONNOLLY, James Brendan, 1868-1957. *The U-Boat Hunters.* NY, Scribner's, 1918. 263 pp., illus. LC 18-13317.

Journalistic anecdotal accounts of anti-submarine warfare. The author was born in South Boston, Mass. and was a champion in the first modern Olympic Games of 1896. Served in the Spanish-American War and was at Santiago in 1898. During 1907-08 was in the USN service. War correspondent for *Collier's* in Mexico in 1914, and in European waters during 1917-18. He was the author of a number of books concerning the sea.

Conquering the Deep Sea Frontier. See entry 284.

The Conquest of the Sea. See entry 1132.

La Conquista de las Profundidades. See entry 1029.

Construction. See entry 464.

Convoy system. See entry 1455.

247 Cook, James Gordon. *Exploring Under the Sea.* Drawings by Leslie Haywood. NY, Abelard-Schuman, 1964. 159 pp. LC 64-14902.

Juvenile.

248 COOKE, David Coxe, 1917- . *How Atomic Submarines are Made.* NY, Dodd, Mead, 1957. 64 pp., illus. LC 57-12133.

Juvenile.

249 ________. *How Atomic Submarines are Made.* Revised ed. NY, Dodd, Mead, 1967. 64 pp., illus. LC 67-7779.

For readers in grades 4-6.

Cookery. See entry 1198.

250 COOMBS, Charles Ira, 1914- . *Deep-Sea World: The Story of Oceanography.* NY, Morrow, 1966. Toronto, McLeod, 1966. 256 pp., illus. LC 66-12610.

Juvenile. Deep Star and other submersibles.

251 COPE, Harley Francis, 1898- . *Serpent of the Seas, The Submarine.* Foreword by Leland P. Lovette. NY and London, Funk and Wagnalls, 1942. xiii, 252 pp., frontis. port., plates. LC 42-50668.

252 ________. *El Submarino, Pasado y Presente.* Buenos Aires, Ediciones Continental, 1944. 229 pp., illus.

Spanish language version of *Serpent of the Seas.*

253 ________ and Walter KARIG, 1898- . *Battle Submerged: Submarine Fighters of World War Two.* NY,

Norton, 1951. 244 pp., illus. LC 51-9593.

Coral Sea, Midway, and Submarine Actions. See entry 884.

254 CORBIN, Thomas W. *The Romance of Submarine Engineering: Containing Interesting Descriptions in Non-technical Language of the Construction of Submarine Boats, the Salving of Great Ships, the Recovery of Sunken Treasures, the Building of Breakwaters and docks, and many other Feats of Engineering Beneath the Surface of the Water.* Philadelphia, Lippincott, 1913. London, Seeley, Service, 1913. 315 pp., illus., diagrams, plates. LC 12-29970.

255 ————. *The Wonders of the Submarine.* London, Seeley, Serivce, 1918. ? pp., illus.

Corey, Herbert, 1872-1954., journalist. *Herbert Corey: A Register of His Papers in the Library of Congress.* Washington, Library of Congress, 1959. 7 pp.

Cork & Orrery, Earl of. See entry 1091.

Cornelissen, Peter, *pseud.* See entry 165.

Cornford, L. Cope. See entry 94.

Costa Brava. See entry 1031.

256 COTTMAN, H. Stuart and Le Vergne SHAW. *Submerged: A Tragedy in One Act.* (Gateway Series of Tested Plays) Evanston, Row, Peterson, 1929.

257 COULTER, Stephen. *Threshold: A Novel.* NY, Morrow, 1964. London, Heineman, 1964. Don Mills, Collins, 1964. 215 pp. LC 64-20296 (64-5021)

The sinking of an Atomic submarine in the Baltic precipitates an underwater U-2 incident.

Course of Lectures Upon the Defense of the Sea-Coast of the United States. See entry 4.

Courten, Raffaele, 1888- . See entry 1166.

258 COUSTEAU, Jacques Yves, 1910- . *The Living Sea* with James Dugan. NY, Harper, 1963. 325 pp., plates, color plates. LC 62-14525.

Same. London, Hamilton, 1963. 212 pp., illus., col. illus.

259 ————. *Le Monde du Silence* with Frédéric Dumas. Paris, Editions de Paris, 1954. 237 pp., illus. LC 54-42021.

260 ________. *The Silent World* with Frédéric Dumas. London, Hamilton, 1953. 148 pp., illus. LC 53-29706.

261 ________. *Silent World.* NY, Pocket Books, 1967. (GC 119)

262 ________. *World Without Sun* edited by James Dugan. NY, Harper, 1965. London, Heineman, 1965. 202 pp., illus. color illus. LC 65-14690.

Cousteau, Jacques Yves, 1910- . See also entries 349, 631, 767, 1032, 1116, 1125, 1419.

263 COUSTEAU, Jacques Yves, 1910- and James DUGAN, editors. *Captain Cousteau's Underwater Treasury.* NY, Harper, 1959. 415 pp.

An anthology by various writers about the underwater world.

264 COWIE, J. S. *Mines, Minelayers and Minelaying.* London, NY, Oxford University Press, 1949. xiii, 216 pp., illus, maps. bibliog. LC 49-48143*.

265 COX, Donald William. *Explorers of the Deep.* Port. illus. by Jack Woodson. (Pioneers of Oceanography) Maplewood, Hammond, 1968. 93 pp., colored maps. LC 68-27451.

For the young reader. Seventeen biographical sketches including: Ben Franklin, Scott Carpenter, and Roger Revelle.

Crabb, Lionel Philip Kenneth. See entries 1002, 1003.

266 CRACKNELL, William H. *USS Barb (SS 220): Gato Class Submarine.* Culver City, Calif., Profile, 1973. 23 pp., illus.

Famous American submarine of WW II.

267 CRAIG, John D., 1903- . *Danger is my Business.* NY, Simon and Schuster, 1938. 309 pp., plates, ports. LC 38-27124.

Memoirs of a professional hard hat diver.

268 ________. *Gefahr ist mein Beruf.* Translated by Karl Eugen Brunner. Zurich, Orell Fussli, 1946. 320 pp., illus., ports. LC 50-22790.

269 ________. *Peligro es mi Profesión.* Translated by Enrique Solorzano. Buenos Aires, Editors del Plata, 1940. 300 pp., plates, ports. LC 48-35162*.

270 CRAIGHILL, William Edward, 1863- . *Foreign Systems of Torpedoes as Compared With Our Own.* Appendix I·The

French Torpedo System . . . by Francis R. Shunk . . . Appendix II. Hasty Notes on Torpedoes by . . . Eric Bergland . . . Willets Point, N.Y.H., Printed on the Battalion Press, Privates Phillips, Kelly and Coventry, printers, 1888. (U.S. Engineer School. Paper no. XI) 20 pp., plates. LC 13-11677.

Crash Dive. See entry 778.

271 THE CRASH DIVE. Hampton Roads, Va., 1920- (v. 1, no. 1—Oct. 2, 1920) "Published at the Submarine Base, Naval Operating Base, Hampton Roads, Va. in the interest of Submarine Division Five and the entire naval service." Newspaper. LC 27-7366.

Creasy, Sir George. See entries 431, 1041, 1045.

272 CRÈPAS, Attilio, 1903- . *Gli Angeli Senza Ali: Vita Segreta dei Sommergibili.* Torino, Paravia, 1939. 119 pp., illus., plates, ports. LC 45-47450.

273 ———. *Vita Segreta dei Sommergibili. . . .* Torino, Paravia, 1943. 128 pp., illus., plates, ports. LC 46-15730.

Cressey, H. M. S. See entry 1441.

274 CRESWELL, John. *Sea Warfare, 1939-45. A Short History.* London and Toronto, Longmans, 1950. xv, 344 pp., maps. LC 50-8364.

Crile, George (Barney), 1907- . See entry 275.

275 CRILE, Jane. *Treasure-Diving Holidays* by Jane and Barney Crile. NY, Viking Press, 1954. 263 pp., illus. LC 54-6426.

The Crisis of the Naval War. See entries 679, 680, 681.

Cristina, Ray. See entry 1123.

276 CROMPTON, Iwan. *U 41; der Zweite Baralong-fall.* Berlin, Scherl, 1917. 116 pp., plates, ports. LC 21-15058.

277 ———. *Englands Verbrechen an U 41, der Zweite Baralong-fall im Weltkrieg.* Edited by W. von Langsdorff. Gutersloh, Bertelsmann, 1941. 224 pp., frontis., plates, ports. LC 45-33778.

Crompton, Iwan. See also entry 110.

278 CROPP, Benjamin, 1936- . *Handbook for Skindivers.* Sydney, Afco Printery, 1963. 147 pp., illus.

279 CROSS, Ellis Royal, 1913- . *Underwater Safety, Including Curriculum for Sport Diving Training Program.* Wilmington, Calif., Diving Research Publishing Co., 1954. 88 pp., illus. LC 54-4129.

280 CROSS, Wilbur. *L'Appel des Profondeurs.* Paris, Presses de la Cite, 1960.

281 ________. *Challengers of the Deep: The Story of Submarines and the Exploits of the Men Who Build and Sail Them.* NY, Sloane, 1959. 258 pp., illus. LC 58-10568.

La Cuestión Submarina Ante el Derecho Internacional. See entry 1148.

D

282 DADIN, Michael. *Submarines.* Chicago, Rand-McNally, 1963. Toronto, Allen, 1963. 96 pp., illus. LC 63-13770.

Juvenile.

Dahlak. See entries 1056, 1957.

A Damned Un-English Weapon. See entry 507.

Danger! And Other Stories. See entry 335.

Le Danger c'est ma Vie. See entry 98.

Danger is My Business. See entry 267.

Danger Zone. See entry 216.

Daniels, Josephus, 1862- . See entries 664, 1290.

283 DANRIT, Captain. *The Sunken Submarine.* Boston, Little, Brown, 1912. 308 pp.

Graphic and technical account of a man trapped in a sunken submarine for a period of six days.

Darby, Patricia. See entry 284.

284 DARBY, Ray and Patricia DARBY. *Conquering the Deep Sea Frontier.* NY, McKay, 1971. 204 pp., illus. LC 74-165082.

Research submersibles. For the younger reader.

Dardanelles Patrol. See entry 1115.

Daring Deeds of Merchant Seamen in the Great War. See entry 1448.

Dark Seas Above. See entry 483.

285 DARRIEUS, Gabriel, 1859- . *Les Sous-Marins.* Paris, Academie de Marine, 1927. 143 pp., plates.

285a ________. *La Guerre Sur Mer, Strategie et Tactique.* Paris, 1907.

This title was translated by Philip R. Alger and published by the U.S. Naval Institute in 1908 with the title *War on the Sea: Strategy and Tactics.*

Daubin, Freeland Allan, 1886- . See entry 1336.

286 DAUDENART, L. G. *La Guerre Sous-Marine et les Torpedoes.* Brussels, Muquardt, 1872. 93 pp., plates.

David Bushnell and his American Turtle. See entry 1255.

287 DAVID, Heather M. *Admiral Rickover and the Nuclear Navy.* NY, Putnam, 1970. 233 pp. LC 78-92807.

Juvenile.

Davidson, Sir Charles Peers, 1841-1929. See entry 187.

288 DAVIS, Sir Robert Henry, 1870-1965, ed. *Deep Diving and Submarine Operations: A Manual for Deep Sea Divers and Compressed Air Workers.* 4th ed. London, The Saint Catherine Press, 1936. 509 pp., frontis., illus., plate, map. LC 36-32894.

289 ________. *Deep Diving and Submarine Operations. . . .* 6th ed. London, Siebe, Gorman, 1955. 603 pp., illus., ports, chart. LC 57-17611.

290 ________. Same. 7th ed. London, St. Catherine Press, 1962. 693 pp., illus.

291 ________. *Deep Diving and Under-Water Rescue.* (Cantor Lectures, 1933) London, Royal Society of Arts, 1934. 62 pp., illus.

292 ________. *A Diving Manual and Handbook of Submarine Appliances.* . . . Embodying the deep Sea Practice adopted by the British Admiralty . . . and including chapters on the physics and physiology of diving, salvage operations, etc., also an account of the evolution of the diving dress and diving bell, and instructions for divers, etc. London, Siebe, Gorman, 1924. 271 pp., illus. LC 25-7845.

Davis, Tenney Lombard, 1890- . See entry 42a.

293 DAWSON, Michael. *Fathoms Deep.* NY, Sheridan House, 1943. 187 pp.

Novel re fictional submarine HMS Rapier.

294 De BERHEGYL, Suzanne. *Ships, Shoals, and Amphoras: The Story of Underwater Archaeology.* NY, Holt, 1961.

Dealey, Sam. See entry 793.

Death of the Thresher. See entry 989.

Decima Flottiglia Mas. See entry 129.

295 De COSTE, Hilaron. *Vie du Pere Mersenne.* Paris, 1649.

Les Découvertes Sous-Marines Modernes. See entry 333.

Deep Challenge. See entry 1186.

Deep Diving and Submarine Operations. See entries 288-291.

Deep Down Under. See entry 409.

Deep is the Blue. See entry 361.

The Deep Range. See entry 226.

Deep Sea Divers. See entry 1494.

Deep-Sea Diving. See entry 842.

Deep-Sea Photography. See entry 597.

Deep Sea Research, Proceedings of the Symposium on Aspects of. See entry 242.

Deep-Sea Salvage. See entry 1456.

Deep-Sea World: The Story of Oceanography. See entry 250.

Deep Star (research submarine). See entry 250.

The Deep Submersible. See entry 1224.

Deep-Water Archaeology. See entry 350.

The Deepest Days. See entry 1180.

Defeat at Sea. See entry 86.

Defensively-Armed Merchant Ships and Submarine Warfare. See entry 608.

296 DEGNAN, James L., Jr. *Submariner!* NY, Dodd Mead, 1959. Toronto, Dodd Mead, 1959. 96 pp., illus. LC 59-6516.

297 DEGOUY, Jean Baptiste Charles Robert Mathieu, 1852- . *Pour en Finir avec les Sous-Marins.* Paris, Payot, 1918. 301 pp. LC 19-7504.

De L'Admirable Pouvoir de L'Art et de las Naturae. See entry 42.

298 DELAGE, Edmond. *La Guerre Sous les Mers.* Paris, Grasset, 1934. 253 pp. LC 34-11643.

299 DELPEUCH, Maurice, 1866- . *La Navigation Sous-Marine à Travers les Siècles.* Paris, Juven, 1902. 450 pp., plates, illus., facsims. LC 2-24568.

An important source book.

300 DEMEOCQ, Andre Joseph Victor, 1907- . *Physiologie de la Plongée. Conférence Faite au Palais de la Découverte le 1er Juin 1950.* Paris, University of Paris, 1951. 20 pp. LC A 54-5855.

Los Demonios del Mar. See entry 447.

De Neuville, Alphonse. See entry 1410.

In Den Tiefen der Meere. See entry 1213.

Denis Papin, Sa Vie et Son Oeuvre. See entry 381.

Denkschrift der Deutschen Regierung. . . . See entry 477.

Denkschrift der Kaiserlich Deutschen Regierung. . . . See entry 478.

301 DENMARK, Krigsministeriet. *Program for Uddannelsen af Mandskabet Ved iste Ingenieurbataillons sømineafdeling.* Copenhagen, 1870. 42 pp. LC CA 18-1763 Unrev'd.

De Secretis Operibus Artis et Naturae. See entry 44.

302 DESSAINT, A. *Navigation Sous-Marine.* Toulon, Bernard, 1892. 63 pp., plates.

Das Deutsche Bild der Russischen und Sowjetischen Marine. See entry 1058.

Das Deutsche Kriegsflotte. . . . See entry 870.

Der Deutsche Kriegsmarine. See entry 795.

Der Deutsche Unterseebootkrieg. See entry 984.

Die Deutschen Kriegsschiffe 1815-1936. See entry 532.

Die Deutschen Kriegsschiffe 1815-1945. See entry 533.

Die Deutschen U-Boote. . . . See entry 469.

Die Deutschen U-Boote: 1906-1945. See entry 600.

Deutschen U-Boote, 1939-1945. See entry 858.

Die Deutschen Waffen und Geheimwaffen. See entry 806.

The Deutschland. See entry 33.

Deutschland (merchant submarine). See entries 143, 714,-717, 734, 750, 1139, 1439.

Deutschlands Hochseeflotte im Weltkrieg. See entry 1096.

Deutschlands U-Boote, der Schrecken Englands. See entry 1110.

The Development of the Lake Type Submarine Torpedo Boats. See entry 743.

Devils of the Deep. See entry 446.

De Vries, W. P. See Vries, W. P. de.

303 DEWITZ, Hrolf von, 1873- . *War's New Weapons. . . .* Introduction by Hudson Maxim. NY, Dodd Mead, 1915. 295 pp., illus.

304 DIBNER, Bern. *The Victoria and the Triton.* (Burndy Library Publication 19) Norwalk, Conn., The Burndy Library, 1962. 58 pp., illus.

Retells and compares the circumnavigations of Magellan in the *Victoria* and Beach in the *Triton.*

305 ________. *The Victoria and the Triton.* NY, Blaisdell, 1962, 1964. 110 pp., illus., bibliog. LC 63-14171.

306 Di CERTO, J. J. *Missile Base Beneath the Sea. The Story of Polaris.* NY, St. Martins, 1967. Toronto, Macmillan, 1967. 165 pp., illus. LC 67-6152.

Grades 7 and up.

A Dictionary of General Dynamics. See entry 472.

Dietz, R. S. See entry 971.

307 DINKLAGE, Ludwig. *U-Boot-Fahrer und Kamelsreiter: Kriegsfahrten eines Deutschen Unterseebootes.* Stuttgart, Franckh'sche, 1939. 212 pp., plates, ports, map. LC AC 40-202.

308 DIOLÉ, Philippe. *L'Aventure Sous-Marine.* Paris, Michel, 1951. 267 pp., illus., bibliog. LC 51-6930.

309 ________. *Four Thousand Years Under the Sea.* (Promenades d'Archeologie Sous-Marine) Translated by Gerard Hopkins. NY, Messner, 1954. 237 pp., illus. LC 54-6762.

310 ________. *The Gates of the Sea.* (Les Portes de la Mer) Translated by Alan Ross. NY, Messner, 1955. 176 pp., illus. LC 55-6708.

311 ________. *Les Portes de la Mer.* Paris, Michel, 1953. 220 pp., illus. LC 53-39813.

312 ________. *The Undersea Adventure.* Translated by Alan Ross. London, Sidgwick & Jackson, 1953. 217 pp., illus. LC 53-36074.

313 ________. *Under-Water Exploration, A History.* Translated by H. M. Burton. London, Elek, 1954. 118pp., illus. LC 55-27709.

Diplomatic Correspondence Between the United States and Belligerent Governments Relating to Neutral Rights and Commerce. See entry 1315.

Diplomatieke Bescheiden Betreffende den Verscherpten Duikbootoorlog. See entry 900.

Disarmament. See entries 531, 539, 540, 1399, 1431-1434, 1436, 1437.

Disasters. See entries 209, 759, 807, 813, 989, 1066, 1110a, 1121, 1122, 1248, 1266, 1308, 1426, 1428.

Discharged Dead. See entry 571.

Discoveries and Inventions of the Nineteenth Century. See entry 1078.

314 DISSETTE, Edward and H. C. ADAMSON. *Guerilla Submarines.* (War Book Original) NY, Ballantine Books, 1972. 238 pp., illus., bibliog. Foreword by F. Kent Loomis. Ballantine Book 416. SBN 345-02597-0-125.

The story of USN submarine supply and rescue missions to the Phillipines from 1 February 1943 through 23 January 1945.

315 DITTMAR, F. J. and J. J. COLLEDGE. *British Warships, 1914-1919.* London, Ian Allen, 1973. illus.

Dive: The Complete Book of Skin Diving. See entries 196, 197.

Dive! The Story of an Atomic Submarine. See entry 570.

Dive in the Sun. See entry 1019.

Divers and Diving. See entry 1457.

Divers in Deep Seas. See entry 843.

Diver's Luck: A Story of Pearling Days. See entry 92.

The Diver's World. See entry 161.

Diving. . . . See entry 629.

Diving. See entries 36, 92, 93, 136, 138, 172, 220, 254, 267, 268, 269, 288-292, 324, 340, 353, 354a, 356, 370, 371, 372, 374, 375, 398, 409, 435, 550, 616, 629, 655, 656, 695, 697, 736, 751, 752, 842, 843, 844, 848, 851, 852, 866, 939, 964, 965, 979, 1036, 1037, 1121, 1122, 1123, 1130, 1132, 1196, 1334, 1372, 1373, 1374, 1376, 1456, 1457, 1467, 1470, 1494, and 1495.

Diving and Salvage. See entry 561.

Diving and Underwater Work. See entry 690.

Diving Bell. See entries 40, 43, 50, 53, 81-84, 107, 127, 244.

Diving, Cutting, and Welding in Underwater Salvage Operations. See entry 1237.

Diving for Science. See entry 991, and 1124.

Diving for Sunken Treasure. See entry 1037.

Diving for Treasure. . . . See entry 1467.

Diving into the Past. See entry 245.

Diving Into the Past. See entry 1007.

Diving Manual. See entry 1367.

Diving Manual, 1924. See entry 1332.

Diving Saucer. See entries 1116, 1125.

Diving With and Without Armor. See entry 525.

Dix Ans et Vingt Jours. See entry 316.

Dix-Neuf Histoires de Sous-Marins. See entry 428.

Doctors of Yale College, 1702-1815. See entry 1241.

316 DOENITZ, Karl, 1891- . *Dix Ans et Vingt Jours.* Paris, Plon, 1959.

317 ———. *Mein Wechselvolles Leben.* (My Fluctuating Life.) Goettingen, Musterschmidt, 1968. 227 pp., illus.

The life and career of former Commander-in-Chief of the German Submarine force.

318 ———. *Memoirs: Ten Years and Twenty Days.* Translated by R. H. Stevens in collaboration with David Woodward. London, Weidenfeld, 1959. NY, World, 1959. 500 pp., illus., maps. LC 59-11530.

See the *Book Review Digest* for 1960 for reviews of this work.

319 ———. *Die U-Bootswaffe.* Berlin, Mittler, 1939. 65 pp., plates. LC A 41-1689.

Important document outlining strategy and tactics for the German U-boat forces just prior to beginning of WW II.

320 ———. *Zehn Jahre und Zwanzig Tage.* (Ten Years and Twenty Days) Bonn, 1958.

"The memoirs of the outstanding figure of the U-boat war. Although slanted to vindicate Doenitz's conduct of the campaign, it is indispensible. The English translation (London, 1959, and NY, n.d.) is inferior to the German original because it omitted the Admiral's voluminous source references and frequently blurred the true meaning of his statements."—Ladislas Farago.

Doenitz, Karl, 1891- . See also entries 431, 432, 434, 573, 698, 729, 791, 802, 923, and 1417.

321 THE DOLPHIN. Groton, Conn., USN Submarine Base, 194?- .

Newspaper for submarine forces afloat and ashore, Atlantic Fleet.

Dolphin Dishes. See entry 1198.

Dolphin, USS (fictional). See entry 821.

322 DOMMETT, W. E. *Submarine Vessels. . . .* London, Whitaker, 1915. 106 pp., illus.

Domville, Barry. See entry 1157.

323 DOMVILLE-FIFE, Charles William, 1886- ed. *Evolution of Sea Power: Studies of Modern Naval Warfare and the Effect of Evolution on the Basis and Employment of Sea Power.* London, Rich & Cowan, 1939. 258 pp., frontis., illus., plates, maps, diagrams. LC 40-7287.

324 ________. *Submarine Engineering of To-Day*. Philadelphia, Lippincott; London, Seeley, Service, 1914. 323 pp., illus., diagrams. LC A 14-375.

325 ________. *Submarine Warfare of Today*. Philadelphia, Lippincott, London, Seeley, Service, 1920. 303 pp., frontis, illus., plates, diagrams. LC 20-26104.

326 ________. *Submarines and Sea Power*. London, Bell, 1919. 250 pp. LC 19-18399.

327 ________. *Submarines, Mines and Torpedoes in the War.* London, NY, Hodder & Stoughton, 1914. 192 pp., frontis., plates, diagram, map. LC 15-5405 Revised.

328 ________. *Submarines of the World's Navies,* with contributions by ADM Sir J. O. Hopkins, ADM Sir Cyprian Bridge, *et al.* London, Griffiths, 1910. 150 pp., frontis, plates, maps, tables. LC 10-27701.

329 DONATO, Alberto. *I Sommergibili Italiani.* 2nd ed.

Rome, Ufficio Storico della Marina Militare, 1968. 310 pp., illus.

Much enlarged over the first edition of 1963.

330 DOOR, W. T., 1914- . *Down in Davey Jones' Locker.* NY, Charlie Noble Press, 1943. 426 pp., illus.

331 DORLING, Henry Taprell, 1883- . *Endless Story: Being an Account of the Work of the Destroyers, Flotilla-Leaders, Torpedo-Boats and Patrol Boats in the Great War.* With a frontispiece by Charles Pears and other illus. by R. T. Amedroz. London, Hodder & Stoughton, 1931. 541 pp., illus., diagrams, maps, bibliog. LC 32-17890.

Written under the pseudonym Taffrail.

332 ________. *Swept Channels.* London, Hodder & Stoughton, 1935. 388 pp. LC 35-22333.

333 DOUKAN, Gilbert. *Les Decouvertes Sous-Marines Modernes.* Preface by Leon Bertin. Paris, Payot, 1954. 329 pp., illus. LC 54-2695.

334 ________. *The World Beneath the Waves.* Translated by A. and R. M. Case. London, Allen & Unwin, 1957. Toronto, Nelson, 1957. 301 pp., illus. LC 57-3522.

Translation of entry 333.

334a ________. *World Beneath the Waves.* Translated by A. and R. M. Case. NY, De Graff, 1957. 356 pp., illus. LC 57-9118.

Down in Davey Jones' Locker. See entry 330.

Down the Hatch. See entry 1477.

Down to the Sea. See entry 1040.

Down to the Sea in Subs. See entry 788.

Down to the Ships in the Sea. See entry 537.

335 DOYLE, A. Conan, 1859-1930. *Danger! and Other Stories.* NY, A. L. Burt, 1911. 310 pp.

Lead story written 18 months before the outbreak of WW I in which German submarine attacks on British and allied merchant shipping are predicted.

The Dragon in the Sea. See entry 592.

Drake, Robert L., *pseud.* See entry 587.

Drama. See entries 256, 696, 801.

Dreadnought ou Submersible? See entry 546.

Dreadnought to Polaris. See entry 651.

336 DREBBEL, Cornelis Jacobsz, 1572-1634. *Grondige Oplossinge, van de Natur en Eygenschappen der Elementen. . . .* Amsterdam, Lamsveld, 1732. 108 pp., plates. LC 15-6271.

337 ________. . . . *Gründliche Aufflosung, von der Natur und Eigenschafft der Elementen. . . .* Frankfurt am Main, Isingin, 1715. 118 pp. LC 15-17628.

338 ________. *De Natura Elementorum,* and *De Quinta Essentia.* Amsterdam?, 1628.

There was a translation into French of these two treatises published in Paris by Jean D'Houry in 1672.

Drebbel, Cornelius van. See Drebbel, Cornelis Jacobsz, 1572-1633.

Drebbel, Cornelis Jacobsz. See also entries 569, 890.

339 DRESSNER, Simon. *Skin Diving.* Chicago, Rand-McNally, 1963. London, Allen, 1963. 95 pp., illus. LC 63-13771.

Juvenile.

340 DRIEBERG, Friedrich Johann, von, 1780-1856. *Mémoire sur une Nouvelle Machine à Plonger, Appelée Triton.* Paris, Didot, 1811. 28 pp., frontis., plates. LC 38-39749.

Drill Book for the Use of Sub-Marine Mining Engineer Volunteers. See entry 114.

Droits et Devoirs du Plongeur Sous-Marin. See entry 351.

341 DROSTE, C. L. and W. H. TANTUM IV. *The Lusitania Case.* Riverside, Conn., 7C's Press, 1973.

Compilation of news articles, editorials, court records, etc.

342 DRUMMOND, John Dorman. *H.M.U-Boat.* London, Allen, 1958. NY, British Book Center, 1958. Toronto, Smithers, 1958. 228 pp., illus. LC 58-4610.

The German U-570 captured by British during WW II was rechristened H.M.S. Graph and put into service by the British.

343 DUEKER, Christopher Wayne. *Medical Aspects of Sport Diving.* Illustrated by J. Edson Adams. NY, A. S. Barnes, 1970. 232 pp., illus., bibliog. LC 69-13035.

Duffner, G. J. See entry 1346.

344 DUGAN, James Thomas, 1912-1968. *Man Explores the Sea.* Penguin Books, 1960. 416 pp., illus., chronology, bibliog. Originally pub. in London by Hamilton in 1956.

345 ________. *Man Under the Sea.* NY, Harper, 1956. 332 pp., illus. LC 55-10692. American ed. of 344 above.

346 ________. *Man Under the Sea.* With a foreword by Jacques Yves Cousteau. New rev. ed. NY, Collier Books, 1965. 443 pp., illus. LC 64-18390.

347 ________. *World Beneath the Sea.* Foreword by Gilbert M. Grosvenor. Washington, National Geographic Society, 1967. 204 pp., illus. LC 67-16081.

Slanted toward a juvenile audience. Superb photography.

348 ________ and Richard VAHAN, eds. *Men Under Water.* (Edited for the Underwater Society of America) Phila., Chilton, 1965. 234 pp., illus., bibliog. LC 65-15943.

349 ________. *Undersea Explorer: Story of Captain Cousteau.* NY, Harper & Row, 1957. London, Hamilton, 1957. 148 pp., illus. (W) 57-9271.

Juvenile biography of Jacques Yves Cousteau.

Dugan, James Thomas, 1912-1968. See also entries 258, 262, 263.

De Duikboot en Haar Rol in den Europeeschen Oorlog. See entry 1418.

Duikboot en Volkenrecht. See entry 427.

350 DUMAS, Frédéric. *Deep-Water Archaeology.* Translated from French by Honor Frost. Chester Springs, PA, Dufour, 1964. London, Routledge, 1962. 71 pp., illus. LC 63-6380.

Dumas, Frédéric. See also entries 259, 260.

351 DUMAS, J. *Droits et Devoirs du Plongeur Sous-Marin.* Paris, Nouvelle Ed. Latines, 1962.

Duncan, Francis. See entry 606.

352 DUNCAN, Ruth H. *The Captain and Submarine CSS Hun-*

ley. Memphis, the author, 1965. 110 pp., illus.

Horace Lawson Hunley and the submarine he commanded at Charleston Harbor during the Civil War.

Duncan-Jones, Arthur Stuart, 1879- . See entry 917.

Du Plat. See Taylor, Joan Du Plat.

353 Du PONT, Alexis Felix, 1879- . *Under Sea With Helmet and Camera. Experiences of an Amateur.* NY, Dodd, Mead, 1940. 87 pp., frontis., plates, port. LC 40-35179.

354 DUPUY, Trevor Nevitt, 1916- . *The Naval War in the West: The Wolf Packs.* NY, Franklin Watts, 1963. London, E. Ward, 1963. 600 pp., illus., maps. LC 63-9797.

Juvenile.

Dust on the Sea. See entry 72.

Dutch Submarines. See entries 241, 1188.

354a Du TEMPLE, M. L., 1819-1889. *Du Scaphandre et de son Emploi a Bord des Navires.* Paris, Arthus Bertrand, 1860. 31 pp., two plates.

About diving and the Cabirol diving suit.

355 DURHAM, Victor G., *pseud.* Wrote the Submarine Boys series which were published at Philadelphia by Altemus between 1909-1912. Six of the original books were reprinted during 1933-1937 in London by Saalfield.

Submarine Boys on Duty: or, Life on a Diving Torpedo Boat.

Submarine Boys Trial Trip; or, Making Good as Young Experts.

Submarine Boys And the Middies; or, Prize Detail at Annapolis.

Submarine Boys Lightning Cruise; or, Young Kings of the Deep.

Submarine Boys For the Flag; or, Deeding their Lives to Uncle Sam.

Submarine Boys and the Smugglers: or, Breaking up the New Jersey Customs Frauds.

Submarine Boys Secret Mission; or, Beating an Ambassador's Game.

Dynamic America. See entry 473.

E

E-11, HMS (British submarine). See entries 190, 191, 1115.

356 EADIE, Thomas. *I Like Diving: A Professional's Story*. With an introduction by RADM Philip Andrews. Boston & NY, Houghton Mifflin, 1929. xiv, 245 pp., frontis, plates, ports. LC 29-11297.

Eadie was a "hard hat" Navy diver who won the Congressional Medal of Honor for his diving work during the salvage of the sunken American submarines S-4 and S-51.

Earth, Sky and Sea. See entry 969.

357 EASTON, Alan Herbert, 1902- . *50 North: An Atlantic Battleground*. Foreword by K. L. Dyer. London, Eyre, 1963. Chester Springs, Pa., Dufour, 1964. Toronto, Ryerson Press, 1966. 287 pp., illus., maps. LC 64-1696.

Eckberg, J. M. See entry 429.

Eclinger, George. See entry 1091.

Edison Storage Battery. See entry 649.

358 EDWARDS, Kenneth. *We Dive at Dawn*. Foreword by Sir Roger Keyes. Chicago, Reilly & Lee, 1941. 412 pp., frontis, plates, ports., maps. LC 41-7673.

British submarines during World War One.

359 ________. *We Dive at Dawn*. Foreword by Sir Roger Keyes. London, Rich & Cowan, 1939. 388 pp., frontis, plates, ports, maps. LC 40-3237.

Edwards, Richard Stanislaus, 1885- . See entry 1377.

Eel, USS (fictional). See entry 73.

Eerden, J. H. M. van der. See entry 1230.

Effect of Time in Submarine Service on Vision. See entry 1342.

360 EHRENKROOK, Friedrich von. *History of Submarine Mining and Torpedoes.* By Captain-Lieut. F. von Ehrenkrook of the Imperial German Navy. Being a supplement to Marine Ordnance Sheet, No. 24 (unofficial) published at Berlin, September 30, 1878. Tr. from the German by Sergeant-Major Frederick Martin, Battalion of Engineers, USA. Willets Point, N.Y.H., Printed on the Battalion Press, Sergt. Carmichael & Pvt. Beck, printers, 1879? 72 pp. (U.S. Engineer School. Professional Papers. Vol. I No. 1) LC 13-11688.

361 EHRLICH, Max. *Deep is the Blue.* NY, Doubleday, 1964. 299 pp. LC 64-16555.

"A solidly plotted novel about the men of a 'jinxed' American nuclear submarine, the USS Cody, and the wives they have left behind. The suspenseful story gains power in the tense atmosphere of command decisions—and the deadly 'rules' of the Cold War game." blurb.

Eight Bells. Royal Mail Lines War Story, 1939-45. See entry 171.

80,000 Milles en Torpilleur. See entry 401.

Einzelkaempfer auf See. See entry 87.

Der Eiserne Seehund. See entry 1229.

ELCO Submarine Chaser. See entry 1171.

362 (ELECTRIC BOAT COMPANY) United States. Electric Boat Company of New Jersey, Committee on (House 60) *Hearings Before United States Select Committee under House Resolution 288.* (Concerning the Lilley and Lake submarine charges.) Parts 1-17. March 9-April 24, 1908. Washington, Government Printing Office, 1908. 17 parts. 1659 pp.

363 ________. *Report of the Select Committee Appointed Pursuant to House Resolution 288.* (Concerning the Lilley and Lake submarine charges.) Washington, 1908. 60th Congress; 1st Session. House Report No. 1727. 3 volumes.

364 ________. 60th Congress; 1st Session. House Report no. 1727.

Part 1. *Report.* 87 pp., 2 facsimiles.
Part 2. Testimony. 1037 pp.
Part 3. Testimony with Index. 870 pp., 22 facsimiles.

The pagination of the three volumes is consecutive. ie. pp. 1-1909.

Electric Boat Company. See also entries 175, 362, 363, 364, 473, 625, 1143, 1314.

Electric Boat Company of New Jersey. See entry 1297.

Elias, Richard A. See entry 1349.

365 ELLACOTT, Samuel Ernest. *Ships Under the Sea.* With drawings by the author. Introduction by Ben Bryant. London, Hutchinson, 1961. 142 pp., illus.

Eller, Ernest M. See entry 131, 367.

Elliott, David Hallen. See entry 95.

366 ELLIS, Frederick D. *The Tragedy of the Lusitania.* Washington, George Berton, Office of the Librarian, 1915. Philadelphia, National Pub. Co., 1915 (Memorial ed.) 324 pp., illus. LC 15-14769.

367 ELLER, Ernest McNeill, 1903- . *The Soviet Sea Challenge.* Foreword by Arleigh Burke. NY, Cowles, 1972. 315 pp., illus., maps. LC 70-163242.

Ellman, Joan. See entry 368.

368 ELLMAN, Van. *The Young Sportsman's Guide to Scuba Diving.* Photographs by Joan Ellman. Nashville, Tenn., Nelson, 1963. 94 pp., illus. LC 63-15586.

369 ELLSBERG, Edward, 1891- . *Hell Below: Photoplay Title of Book Pigboats.* (Illus. with scenes from the movie) NY, Grosset, 1933. 329 pp.

370 ________. *Men Under the Sea.* NY, Dodd, Mead, 1939. 365 pp., frontis., plates, ports, facsimiles. LC 39-27859.

Hard hat diving anecdotes and stories of the salvage of the USS S-4 (SS 109), S-51 (SS 162), and *Squalus* (SS 192).

371 ________. *On the Bottom.* NY, Dodd, Mead, 1929. 324 pp., frontis., plates, ports. Illus. end papers. LC 29-11296.

An account of the raising of the USS S-51 (SS 162).

372 ________. *On the Bottom: An Epic of Deep Sea Diving.* Edited and with an introduction by Hector C. Bywater. London, Constable, 1929 and 1933. 307 pp., illus.

373 ________. *Pigboats.* NY, Dodd, Mead, 1931. 329 pp. LC 31-6275.

The movie *Hell Below* was based on this book.

374 ________. *Report on Salvage Operations: Submarine S-51.* Washington, Government Printing Office, ca. 1927. 1927.

375 ________. *S-51: Par le Fond.* Translated by Gilberte Marchegay. Preface by Rene Jouan. Paris, Arthaud, 1952. 286 pp., illus. LC 52-34670.

Translation of *On The Bottom.*

376 ________. *S-54: Stories of the Sea.* NY, Dodd, Mead, 1932. 278 pp. LC 32-5749.

377 ________. *Submerged: An Epic of the Submarine Service.* London, Hurst, 1934. 286 pp.

Reprint of *On The Bottom.*

Ellsberg, Edward, 1891- . American engineer, inventor and writer. Graduate of the US Naval Academy, class of 1914. Rose through the ranks to Captain. He was salvage officer during the raising of the S-4, S-51 and the Squalus. Served on active duty during both World Wars. See also entry 1419.

Els Homes-Peixos. See entry 1033.

Ely, George H. See entry 1191.

Eminent Americans. See entry 1035.

Employment of Submarines. See entry 1330.

End, Edgar. See entry 1039.

Endless Story. See entry 331.

The Enemy. See entry 1469.

The Enemy Below. See entry 1015.

Enemy Sighted. See entry 638.

Enemy Submarine. See entry 430.

The Enemy Submarine. See entry 1324.

Englands Verbrechen an U41. . . . See entry 277.

English submarines. See entries 134, 135.

L'Enigme des Sous-Marins Soviétiques. See entry 635.

Enriques, Enrique. See entry 170.

Die Enstehungsgeschichte der U-Boote. See entry 760.

Die Entwicklung des U-Bootkrieges. . . . See entry 480.

Environmental Effects on Consciousness. See entry 1092.

Epics of Salvage. See entry 844.

L'Epopee du Surcouf et le Commandant Louis Blaison. See entry 544.

378 *Epron.* No. 1- . 1933- . Leningrad. Russian professional journal re submarine diving and salvage. 12 issues were published between 1933 and 1935. There was an index with issue No. 12. LC 50-42517.

Epstein, Beryl Williams. See entry 379.

379 EPSTEIN, Samuel, 1909- . *The Real Book About Submarines.* By Samuel Epstein and Beryl Williams. Illus. by Manning de V. Lee. Garden City, NY, Garden City Books, 1954. (by arrangement with F. Watts.) 222 pp., illus. LC 54-5474.

________. ________. revised ed. 1962. 216 pp. LC 62-7626.

380 EQUEVILLEY, Raymond d', 1873- . *Les Bateaux Sous-Marins et les Submersibles.* Paris, Gauthier-Villars, 1902. 164 pp., illus. LC 2-28446.

381 ERNOUF, Alfred Auguste, 1817-1889. *Denis Papin, Sa Vie et Son Oeuvre.* Paris, 1874.

Eron, Leonard D. See entry 1340.

Das Erste Handels-Unterseeboot Deutschland. . . . See entry 1439.

Escape. See entries 1338, 1339, 1346.

Essai sur l'Histoire et la Situation du Sous-Marin en Droit International. See entry 1263.

382 ESTRANY, Jeronimo. *Narciso Monturiol y la Navegacion Submarina.* Juicios criticos emitidos sobre los importantisimos Trabajos Realizados por este Sabio Inventor Catalan. BARCELONA, GILI, 1915. 152 pp., frontis., illus., ports, facsimiles. LC 17-9648.

About Narcisco Monturiol, 1819-1885, and his submarine inventions.

Evans, E. R. G. R. See entry 41.

383 EVENING MAIL, NEW YORK. *The Gravest 366 Days.* Editorials reprinted from the New York Evening Mail. NY, 1916. 622 p. LC 16-22914.

Event 1000: A Novel. See entry 759.

384 EVERITT, Don. *The K Boats: A Dramatic First Report on the World's Most Calamitous Submarines.* NY, Holt, 1963. 206 pp., illus. LC 63-18427.

Same. With subtitle: *A Dramatic First Report on the Navy's Most Calamitous Submarines.* London, Harrap; Toronto, Clarke-Irwin, 1963. 206 pp., illus., plates. LC 64-2534.

Evolution of Sea Power. See entry 323.

The Evolution of the Submarine. . . . See entry 1137.

The Evolution of the Submarine Boat. . . . See entry 1205.

385 EWART, William Dunlop. *Atomic Submarines.* With drawings by K. M. Sibley. (Sportshelf Book) London, Weidenfeld, 1963. New Rochelle, Soccer, 1963. 143 pp., illus. LC 64-5663.

Ewing, Maurice. See entry 1419.

Experimental submarines. See entries 1285, 1291, 1298, 1309.

La Exploración Submarina. See entry 1030.

Exploration Sous-Marine. See entry 1017.

Exploration Sous-Marine de la Bible. See entry 411.

Explorers of the Deep. See entry 265.

Exploring the Ocean Depths. See entry 1125.

Exploring the Ocean World. See entry 657.

Exploring the Sky and Sea. See entry 825.

Exporing the World of Oceanography. See entry 1223.

Exploring Under the Sea. See entry 247.

Exploring Under the Sea. See entry 901.

F

Die Fahrt der Deutschland. See entries 714, 715.

386 FAIRHALL, David. *Russian Sea Power.* Boston, Gambit, 1971. 286 pp., illus. ISBN 0-87645-040-0 LC 71-118209.

387 FALCON-BARKER, Ted. *Roman Galley Beneath the Sea.* Philadelphia, Chilton Book Co., 1967. (1st ed. Leicester, Brock-Hampton Press, 1964) 121 pp., illus. LC 67-22177.
Underwater archaeology.

388 ________. *1600 Years Under the Sea.* NY, McKay, 1960. 225 pp., illus. LC 61-7990.
Underwater archaeology.

Famous Underwater Adventures. See entry 1419.

389 FANE, Francis Douglas. *The Naked Warriors,* with Don Moore. Foreword by Richmond Kelly Turner. NY, Appleton-Century-Crofts, 1956. 308 pp., illus. LC 56-10469.
Underwater Demolition Teams, WW II.

390 FANE, Robert, *pseud. Ships May Proceed.* London, Allen, 1943. 63 pp. LC 44-6235.
UDT, WW II

391 ________. *We Clear the Way.* London, Allen, 1943. 64 pp. LC 43-17286.
UDT, WW II

The Far and the Deep. See entry 1169.

392 FARAGO, Ladislas, 1906- . *The Tenth Fleet.* NY, Obolensky, 1962. Toronto, McLeod, 1962. 366 pp., bibliography, notes. LC 62-18782.
WW II U-boat warefare. The story of submarines and survival. Extensive critically annotated bibliography.

Faran är Mitt Liv. See entry 99.

Farrère, Claude. See entry 428.

Fatal Submarine Accidents. See entry 1066.

Fathoms Below. See entry 851.

Fathoms Deep. See entry 293.

393 FAY, Harold J. W. *History and Development of Submarine Signals.* Boston, 1912. 17 pp., illus. plates, diagrams. LC 13-19101.

A paper presented at the 29th annual convention of the American Institute of Electrical Engineers. Boston, June 27, 1912.

394 FAYLE, Charles Ernest, 1881- . *Seaborne Trade* (History of the Great War). NY, Longmans, 1920-24. London, Murray, 1920-24. LC 21-3499.

vol. I. *The Cruiser Period.* (1920)
vol. II. *From the Opening of the Submarine Campaign to the Appointment of the Shipping Controller.* (1921)
vol. III. *The Period of Unrestricted Submarine Warfare.* (1924)

395 FECHTER, Hans. *In der Alarmkoje von U35.* Berlin-Wien, Ullstein, 1918. 151 pp. LC 20-11295.

Feind im Fadenkreuz. See entry 573.

396 FELL, William Richmond. *The Sea Our Shield.* London, Cassell, 1966. Toronto, Longmans, 1966. 232 pp., plates, maps. LC 66-78859.

Midget submarines, WW II.

397 FELSEN, Henry Gregor, 1916- . *He's In Submarines Now.* With pictures by USN and Martin Sheridan. NY, McBride, 1942. 175 pp., illus. LC 42-50556.

Submarine school and training at USN Submarine Base, New London.

398 ________. *Navy Diver.* NY, Dutton, 1942. 233 pp. LC 42-19934.

399 ________. *Submarine Sailor.* NY, Dutton, 1943. 208 pp. plus 32 pp. pictorial supplement of submarine photos. LC 43-11138.

Felsen, Gregor or Henry Gregor is *pseud.* of Henry Vicar.

400 FENDRICH, Anton. *An Bord: Kriegserlebnisse bei den See-und Luftflotten.* Stuttgart, Franckh'lche, 1916. 140 pp.

401 FERRIERE, Maxime Robert Marie, 1893- . *80,000 Milles en Torpilleur: Recits de Chasse aux Sous-Marins. (1914-1916).* Paris, Perrin, 1918. 297 pp. LC 25-15314.

402 FIELD, Adelaide Anderson. *Auguste Piccard, Captain of Space, Admiral of the Abyss.* NY, Houghton, 1969. 150 pp., illus. LC 73-86299.

For juvenile audience.

403 ________. *Challenge of the Sea Floor.* NY, Houghton Mifflin, 1970. 133 pp., illus. ISBN 0-395-10913-2. LC 75-122909.

For juvenile audience. Research submersible *Alvin.*

404 ________. *Man's Rush to the Seafloor.* NY, Houghton Mifflin, 1970.

Underwater exploration.

405 FIELD, Cyril, 1859- . *The Story of the Submarine from the Earliest Ages to the Present Day.* Illus. by the author. London, Low, Marston, 1908. 304 pp., frontis, illus. LC W 9-102.

50 North: An Atlantic Battleground. See entry 357.

The Fight for the Sea. See entry 777.

The Fighting Fleets. See entry 947.

Fighting Ships of Australia and New Zealand. See entry 24.

Fighting Ships of the Imperial Japanese Navy. See entry 450.

Fighting the U-Boats. See entry 217.

Fighting Under the Sea. See entries 815, 816.

406 FINNEY, Jack. *Assault on a Queen.* NY, Simon & Schuster, 1959. 244 pp. LC 59-11194.

Suspenseful novel re salvaged German U-19 (fictional WW I U-boat) and attack and robbery of the SS Queen Mary. Made into a successful movie starring Frank Sinatra.

407 FIORAVANZO, Giuseppe, ed. *Sommergibili in Guerra. Ufficio Storica della Marina Militare.* n.p., Ministero della Difesa-Marina, 1956. 216 pp., tables.

Fire in Anger. See entry 833.

Fire on the Beaches. See entry 1220.

The First Book of Submarines. See entry 652.

First Under the North Pole. The Voyage of the Nautilus. See entry 20.

408 FITZGERALD, Charles Cooper Penrose, 1841-1921. "Submarine Warfare." London, 1919. An article which appeared in the *National Review* vol. 72: pp. 750-755.

Fitzgerald, Edward. See entries 153, 154, 751, 752, 779.

Flasher, USS (SS 249). See entries 527, 528.

Fleet Submarines. See entry 1287.

Fleet Submarines. See entry 1306.

The Fleet Type Submarine. See entry 1355.

409 FLOHERTY, John Joseph, 1882- . *Deep Down Under.* Philadelphia, Lippincott, 1953. 146 pp., illus. LC 53-8912.

Book on diving in "Career Books for Teen-Agers" series.

Floyd, Ronald J. See entry 1007.

Flot Nashei Rodiny. See entry 723.

Flot v Pervoi Mirovoi Voine. See entry 953.

Flying, Submarining and Mine Sweeping. See entry 17.

Flying Fish (fictional submarine). See entry 239.

The Flying Sub. See entry 1145.

410 FOCK, Harald. *Marinekleinkampfmittel.* Munich, Lehmans, 1968. 151 pp., illus., diagrams.

A technical treatise on German midget submarines, explosive boats and frogmen.

411 FÖEX, Jean Albert. *Exploration Sous-Marine de la Bible.* Photos by Serge de Sazo. Paris, France-Empire, 1955. 286 pp., illus. LC 57-22029.

Deals with the 1955 Jonas expedition in Palestine.

412 FÖEX, G. A. *Histoire Sous-Marine des Hommes.* Paris, Laffont, 1964.

Fontes. See entry 960.

413 FONTIN, Paul J. *Les Sous-Marine et L'Angleterre.* Paris, Chapelot, 1902. 70 pp.

Fontin, Paul. See entry 921.

Foreign Systems of Torpedoes. . . . See entry 270.

414 FOREST, Fernand, 1851-1914. *Les Bateaux Sous-Marins. . . .* with H. Noalhat. Preface by A. Saissy. Paris, Dunod, 1900. Two volumes. LC 2-24884.

Volume one deals with history, and volume two with technology. Many illustrations including portraits and diagrams. An important source work on history of submarines.

415 FORSTMANN, Walter. *U39 auf Jagd im Mittelmeer.* Foreword by ADM. Hollweg. Berlin-Wien, Ullstein, 1918. 180 pp. LC 34-3376.

416 FORSTNER, Georg Günther, *Freiherr* von, 1882-1940. *Als U-Boots-Kommandant Gegen England.* Berlin-Wien, Ullstein, 1916. 221 pp. LC 23-4671.

417 ________. *The Journal of Submarine Commander von Forstner.* Translated by Mrs. Russell Codman, with introduction by John Hays Hammond, Jr. Boston & NY, Houghton Mifflin, 1917. liv, 135 pp., frontis., plates. LC 18-8.

418 ________. *Das Kampfschicksal der Deutschen Flotte im Weltkrieg.* Leipzig, Seemann, 1937. 423 pp., illus., maps, plates, ports, facsims. LC 38-17597.

419 ________. *U-Boot vor dem Feind.* Berlin, Weise, 1939. 64 pp., illus. LC 42-48747.

420 ________. *U-Boots-Kommandanten und Kriegsverbrecher: Unbekanntes von unseren U-Booten.* Berlin, Frundsberg, 1936. 297 pp., illus. LC 37-4966.

421 ________. *U-Boots-Leben.* Illus. with reproductions of paintings by C. Bossenroth. Bielefeld und Leipzig, Velhagen and Klasing, 1916?. 32 pp., illus. LC 19-9164.

Forty Fathoms Deep. See entries 655, 656.

Foster, Robert J. See entry 1345.

422 FOUCHER-CRÉTEAU, Jean, 1915- . *Perles, Requins et Coraux.* Paris, Librairie des Champs-Elysées, 1955. 54 pp., 87 plates, partly colored. Captions in French and English. LC 55-37665.

Red Sea diving expedition to investigate marine fauna.

Four Thousand Years Under the Sea. See entry 309.

423 FOURNIER, Georges, 1595-1652. *Hydrographie: Contenant la Theorie et la Pratique de Toutes les Parties de la Navigation.* Paris, Du Puis, 1667. 708 pp., illus. LC F-2905.

424 FRADIN, Brize. *Chimie Pneumatique Applique aux Travaux Sous L'Eau.* Paris, 1808.

Die Frage der Minen im Seekrieg. . . . See entries 1050, 1051.

425 FRANCE. MARINE. GROUPE D'ETUDES ET DE RECHERCHES SOUS-MARINES. *The Complete Manual of Free Diving* by Philippe Tailliez and others of the Undersea Research and Development Group of the French Navy. NY, Putnam, 1957. 185 pp., illus. LC 57-9524.

The original book *Plongée en Scaphandre* (*Diving in Diving Dress*) was published in 1949 by Editions Elzévir.

426 ________. *La Plongée,* by Philippe Tailliez and others. Grenoble, Arthaud, 1955. 154 pp., illus., port, bibliog. LC 55-43659.

France and the Submarines. See entry 1435.

La France et les Sous-Marins. See entry 1435.

427 FRANÇOIS, Jean Pierre Adrien, 1889- . *Duikboot en Volkenrecht.* 's-Grăvenhage, Nijhoff, 1919. 375 pp., bibliog. LC 20-6955 rev.

Considerations of submarine warfare and international law.

428 FRANK, Bernard, 1887- . *Dix-Neuf Histoires de Sous-Marins.* Preface by Claude Farrère. Paris, Payot, 1918. 233 pp. LC 19-2266.

429 FRANK, Gerold, 1907- and James David HORAN, 1914- *U.S.S. Seawolf: Submarine Raider of the Pacific.* As told by J. M. Eckberg. NY, Putnam, 1945. Toronto, Allen, 1945. 197 pp., illus. LC 45-7140.

430 FRANK, Wolfgang, 1909- . *Enemy Submarine: The Story of Gunther Prien, Captain of U-47. Written from Prien's Diaries and from Sailing with him in Action.* Translated by Lynton Hudson. London, Kimber, 1954. 200 pp., plates. LC 56-2967.

431 ________. *The Sea Wolves. The Story of the German*

U-Boats at War. Foreword by ADM Sir George Creasy. London, Weidenfeld & Nicolson, 1955. Translated by R. O. B. Long. 252 pp., illus., diagrams of U-boat.

Admiral Doenitz and the German U-boats of WW II. Translation of *Die Wolfe und der Admiral.*

432 ________. *The Sea Wolves. The Story of German U-Boats at War.* Translated by R. O. B. Long. NY, Rinehart, 1955. 340 pp. LC 55-6425.

433 ________. *Der Stier von Scapa Flow* (The Bull of Scapa Flow) Hamburg, Stalling, 1958. 292 pp., illus.

Gunther Prien and his penetration of British defenses of Scapa Flow in U-47.

434 ________. *Die Wölfe und der Admiral.* Oldenburg, Stalling, 1955. 580 pp., illus.

435 FRANZÉN, Anders. *Warship Vasa: Deep Diving and Marine Archaeology in Stockholm.* Stockholm, Norstedt & Bonnier, 1960.

436 FRASER, Ian Edward. *Frogman V.C.* London, Angus and Robertson, 1957. 216 pp.

UDT in WW II.

437 FRAZAR, Douglas. *Perseverance Island, or The Robinson Crusoe of the Nineteenth Century.* Boston, Lee and Shephard, 1884. 373 pp., illus.

Story of a loner who led a Crusoe-like existence on a South Pacific island. A submarine, powered by goats on a treadmill, was used to explore for sunken treasure.

Free Diving. See entry 1018.

Free Escape from Submarines. See entry 1338.

Freeman, Fred. See entry 1071.

438 FREIRE, Theotonio. *Navios Perdidos.* Porto Alegre, Barcellos. Bertaso & Co., 1937. 176 pp. LC 43-20909.

439 FREIWALD, Ludwig, 1898- . *Last Days of the German Fleet.* Translated by Martin Moore. Introduction by Hector C. Bywater. London, Constable, 1932. 318 pp., maps. LC 33-1434.

440 ________. *Schlachtkreuzer im Nebel.* Munich, Lehmann, 1934. 70 pp. LC 34-40837.

441 ________. *U-Boots-Maschinist Fritz Kasten: Das Frontbuch der Deutschen Kriegsmarine.* Munich, Lehmann, 1933. 327 pp. LC 34-0241.

442 ________. *Die Verratene Flotte; aus den Letsten Tagen der Deutschen Kriegsmarine.* Munich, Lehmann, 1937. (c. 1931) 297 pp. LC 50-50956.

Fremantle, Edmund. See entries 462, 463.

French Merchant Marine. See entry 738.

French Navy. See entries 685, 766, 891.

French Submarines. See entries 105, 132, 133, 206, 210, 228, 229, 544, 764, 765, 779, 1067, 1076, 1236, 1408, and 1435.

443 FREY, Hank and Shaney. *130 Feet Down, Handbook for Hydronauts.* NY, Harcourt, 1967.

444 FREYER, Paul Herbert. *Der Tod auf Allen Meeren.* (Death on all the Oceans). Berlin, Militär Verlag der Deutschen Demokratischen Republik, 1972. 390 pp., illus.
East German viewpoint of WW II German U-Boat warfare.

Frogman: Commander Crabb's Story. See entry 1003.

Frogman V. C. See entry 436.

The Frogmen. See entry 1422.

Frogmen: Training, Equipment and Operations of our Navy's Undersea Fighters. See entry 234.

Frogmen. See entries 87, 88, 105a, 609, 827, 943, 1002, 1003.

From 1900 Onward. See entry 41.

From Pulpit to Prison. See entry 916.

From Sails to Atoms. See entry 992.

From the Turtle to the Nautilus. See entry 633.

From U-Boat to Pulpit. See entries 915, 916.

Frømandsfaerd. See entry 1275.

445 FROST, Honor. *Under the Mediterranean: Marine Antiquities.* NY, Prentice-Hall, 1963. London, Routledge, 1963. 278 pp., illus. LC 63-9710 (63-25317)
Underwater archaeology.

Frost, Honor. See also entry 350.

446 FROST, Wesley, 1884- . *Devils of the Deep*. London, The Field and Queen, 1918? 8 pp. LC 27-782.

447 ________. *Los Demonios del Mar*. Londres, The Field & Queen, (Horace Cox) Limited, 1918. 8 pp. LC 27-22070.

448 ________. *German Submarine Warfare. A Study of its Methods and Spirit, Including the Crime of the Lusitania*. A record of observations and evidence. Introduction by Frank Lyon Polk. NY, Appleton, 1918. xv, 243 pp., illus. LC 18-23030.

449 ________. *La Guerra Submarina de Alemania. Verdadera cronica de Algunos Procederes que Motivaron la Entrada de los Estados Unidos en la Guerra Mundial. . . .* Carcas, Tip. Americana, 1917? 13 pp., illus. LC 19-3661.

450 FUKUI, Shizuo. *Fighting Ships of the Imperial Japanese Navy*. Tokyo, 1970. 327 pp., photos, diagrams, Oblong format. Technical data.

A sort of Japanese "Jane's *Fighting Ships*." Hundreds of photos, many never before published. An excellent 28 pp. section on submarines with many full page photos of submarines under construction or underway. The text is in Japanese.

451 ________. *The Japanese Navy At the End of World War Two*. Old Greenwich, Conn., WE, Inc., 1970. 175 pp., illus., diagrams. Comprehensive coverage on submarines. SBN 991964-000-17.

Fukui was an officer in the IJN during WW II, and is now a LCDR in the Japanese Self Defense Force, and works as an engineer at Mitsubishi. Researcher and historian of Japanese submarines.

Full Crash Dive. See entry 130.

452 FULTON, Robert, 1765-1815. *De La Machine Infernale Maritime*. Paris, 1812. 61 pp., plate.

A French edition of *Torpedo War and Submarine Explosions*.

453 ________. *Letter to the Secretary of the Navy . . . On the Practical Use of the Torpedo*. Washington, 1811. 55 pp.

A later edition of his *Torpedo War . . .*

454 ________. *Letters . . . on Sub-Marine Navigation and Attack. . . .* London, 1806. 37 pp.

455 ________. *Torpedo War and Submarine Explosions.* . . . NY, W. Eliot, 1810. NY, W. Abbott, 1914. 55 pp., illus.

456 ________. *Torpedo War and Submarine Explosions.* (Crerar Classics Series) Chicago, Swallow, 1971. ISBN 0-8040-0533-8.

A reprint of the 1810 edition.

457 ________. *Torpedo War and Submarine Explosions.* NY, Printed by W. W. Elliot, 1810. 57 (3) pp. 5 plates. LC 8-34927.

458 ________. *Torpedo War and Submarine Explosions.* NY, Ptd. by W. Eliot, 1810. Reprinted: NY, W. Abbott, 1914. 55 pp., 5 plates. (On cover: *The Magazine of History.* Extra no. No. 35) LC 15-6380.

459 ________. *De La Machine Infernale Maritime.* Translated by E. Nunez de Taboada. Paris, 1812.

FULTON, Robert 1765-1815.

The bibliographical information on Fulton and his works is a little confusing. Some of the titles listed above may be the same books. There are a number of letters and reports concerning Fulton's American torpedo experiments and proposed system of coastal defense which may be found in *American State Papers. Naval Affairs.*

There are a number of books dealing with Fulton, his life and his work with steam navigation and submarines. These include:

Dickinson, H. W.—*Robert Fulton, Engineer and Artist.* (1913)
McFee, I. N.—*Story of Robert Fulton.* (1922)
NY Public Library—*List of Books, Prints, Manuscripts, etc. Relating to the Hudson, Robert Fulton and Steam Navigation.* (NYPL, 1909)
Parsons, W. B.—*Robert Fulton and the Submarine.* See entry 951.
Perry, F. M.—*Four American Inventors.* (1901)
Sutcliffe, A. C.—*Robert Fulton and the Clermont.* (1909)
Sutcliffe, A. C.—*Robert Fulton.* (1915)
Thurston, R. H.—*Robert Fulton.* (1891)

See also entries 237, 951, 1419.

460 FÜRBRINGER, Werner. *Alarm! Tauchen!! U-Boot im Kampf und Sturm.* Berlin, Ullstein, 1933. 257 pp., plates, ports. LC 34-10218.

Furreg, Herbert. See entry 1213.

Furreg, Waltraut. See entry 1213.

461 FUSS, Richard. *Der U-Boot-Krieg des Jahres 1915. Ein Kapitel Auswärtiger Politik im Weitkriege.* Stuttgart, Kohlhammer, 1936. 96 pp., LC 37-7394.

The Future of the Sea-Based Deterrent. See entry 1270

Fye, Paul M. See entry 496.

462 FYFE, Herbert C. *Submarine Warfare: Past, Present, and Future.* Introduction by Edmund Fremantle. Chapter on submarine construction by Sir Edward J. Reed. NY, Dutton, 1902. London, Richards, 1902. xxviii, 332 pp., illus., bibliog. Appendices include a 12 pp. section on the submarines of Simon Lake.

The author was the Librarian of the Royal Institution in London. This is an important source book.

463 ________. *Submarine Warfare, Past and Present.* Introduction by ADM Sir Edmund Fremantle. 2nd edition, revised by John Leyland. London, Richards, 1907. xxviii, 302 pp., frontis., illus., plates, bibliog. LC 7-32863.

G

464 GABLER, Ulrich. *Unterseebootsbau.* Darmstadt, Weher & Wissen, 1964. 156 pp., illus., tables.

Covers all aspects of modern, non-nuclear, submarine construction. An excellent treatise.

465 GAGET, Maurice. *La Navigation Sous-Marine: Généralités et Historique.* Paris, Beranger, 1901. 472 pp., illus., diagrams, bibliog. LC 2-7071.

466 GALLAGHER, Thomas Michael, 1918- . *The X-Craft Raid.* NY, Pinnacle Books, 1972. 186 pp.

467 GALLERY, Daniel Vincent, 1901- . *Twenty Million Tons Under the Sea.* Chicago, Regnery, 1956. 344 pp., illus. LC 56-11853.

Gallery was in charge of USN task force which took U-505 as a prize ship. Gallery and the task force received the Presidential Unit Citation for the capture. The submarine is now a permanent exhibit of the Museum of Science and Industry in Chicago. Published in England under title *We Captured a U-Boat.*

468 ________. *We Captured a U-Boat.* Introduction by Edward Young. London, Sidgwick, 1957. 268 pp., plates, maps.

Gallery, Daniel Vincent, 1901- . See also entry 888.

Garnett, Richard. See entry 541.

The Gates of the Sea. See entry 310.

469 GAYER, Albert. *Die Deutschen U-Boote in Ihrer Kriegführung, 1914-1918.* Berlin, Mittler, 1920-21. 3 volumes in 1. LC 35-21422. There was also a 4 volume edition with the last volume being published in 1930.

Contents: (1) Von kriegsanfang bis februar 1915. (2) Die U-bootsblockade, februar bis oktober 1915. (3) Der winter 1915/16 und die zeit bis zur anordnung des reinen U-kreuzerkrieges, oktober 1915 bis ende april 1916.

Geer, Charles. See entry 633.

Gefahr ist mein Beruf. See entry 268.

Das Geheimnis um U-110. See entry 1074.

Geheimnis um U-977. See entry 1090.

Das Geister U-Boot . . . See entry 737.

470 GELFOND, Grigory Mikhailovich. *Sovetsky Flot v Voine s Yaponiei.* (The Soviet Fleet in the War with Japan.) Moscow, 1958.

471 GELOW-GERMAN, K. E. *U-Boot Abenteuer im Operrgebiet.* Berlin, Scherl, 1917. 112 pp.

472 GENERAL DYNAMICS. *A Dictionary of General Dynamics.* NY, General Dynamics, 1967. 63 pp., illus.

473 GENERAL DYNAMICS CORPORATION. *Dynamic America: A History of General Dynamics Corporation and its Predecessor Companies,* edited by John Niven, Courtlandt Canby, and Vernon Welsh. Designer Erik Nitsche. NY, Doubleday and General Dynamics, 1960. 426 pp., illus., ports. LC 58-13905.

474 GENERAL DYNAMICS CORPORATION. ELECTRIC BOAT DIVISION. *Submarine Tankers: Summary of a Feasibility Report.* Groton, Conn., General Dynamics Corp., 1959. 22 pp., illus.

475 GENERAL DYNAMICS CORPORATION. ELECTRIC BOAT DIVISION. THE SUBMARINE LIBRARY. *United States Submarine Data.* 3rd ed. Groton, Conn., The Submarine Library, 1958. 23 pp.

Lists, in tabular form, the hull number, name, design, shipyard, launch date, sponsor, first commissioning date, first commanding officer, and final disposition of American submarines from the Holland (SS 1) to 1958. Supplements have been issued to bring this up-to-date.

476 GENOVA TORRUELLA, Arturo, 1889- . *Lanzamiento Nocturno Presuroso.* Madrid, Escelicer, 1942. 46 pp., tables, diagrams. LC 43-18243.

German, Italian, Japanese U-Boat Casualties. See entry 512.

The German Pirate. See entry 11.

The German Navy in WW II. See entries 1416, 1417.

German Submarine Activities on the Atlantic Coast of the United States and Canada. See entry 1385.

The German Submarine War, 1914-1918. See entry 484.

German Submarine Warfare. See entry 448.

German Submarines. See entry 773.

German Submarines. See entries 7, 11, 12, 13, 30, 31, 33, 37, 39, 55, 56, 62, 65-68, 77, 79, 80, 85-91, 106, 109, 110, 120, 126, 143, 144, 159, 164-169, 173, 182, 185, 198, 201, 202, 203, 207, 213, 214, 216-219, 222, 223, 224, 246, 276, 277, 307, 316-320, 342, 354, 366, 394, 395, 400, 401, 410, 415-421, 427, 428, 430-434, 439-442, 444, 446-449, 460, 461, 467, 468, 469, 471, 477-480, 482, 484, 488, 492, 502, 505, 506, 508, 509, 513-519, 521, 531, 535, 536, 543, 562, 573-577, 591, 594, 598, 599, 600-604, 608, 614, 615, 620, 624, 634, 636, 645-648, 664, 671, 672, 679, 680-684, 693, 705, 706, 707, 711, 713-717, 727, 730, 733, 734, 735, 737, 738, 748, 750, 753, 756, 757, 758, 775, 786, 795, 804, 805, 806, 819, 826, 837, 839, 845, 846, 856, 857, 888, 902-905, 910, 911, 915-918, 922, 923, 928, 929, 941, 942, 954, 955, 956, 978, 984, 998, 999, 1011, 1020, 1023, 1024, 1034, 1038, 1041-1046, 1048, 1053, 1053a, 1054, 1055, 1061, 1062, 1072, 1073,

1087, 1090, 1091, 1091a, 1096, 1097, 1098, 1103, 1104, 1105, 1107-1110, 1113, 1114, 1135, 1136, 1139, 1156-1163, 1165-67, 1178, 1189, 1221, 1228, 1229, 1230, 1232-35, 1273, 1274, 1315, 1316, 1321, 1324, 1383, 1384, 1385, 1396-98, 1400-1406, 1411, 1418, 1421, 1440, 1441, 1444, 1448-50, 1459-61, 1472.

German Subs in Yankee Waters. See entry 671.

Germaniawerft. See entry 1221.

Germanskaia Podvodnaia Voina 1914-1918 gg. See entry 485.

477 GERMANY. *Auswärtiges amt. Denkschrift der Deutschen Regierung über die Ermordung der Besatzung eines Deutschen Unterseeboots Durch den Kommandanten des Britischen Hilfkreuzers Baralong. . . .* Berlin, 1915. 43 pp. LC 26-19313.

478 ________. *Denkschrift der Kaiserlich Deutschen Regierung über die Behandlung Bewaffneter Kauffahrteischiffe. . . .* Berlin, 1916. 45 pp. LC 17-11628.

479 ________. *Der Baralong-Fall.* Berlin, Reichsdruckerei, 1916. 64 pp. LC 20-13049.

480 GERMANY, KRIEGSPRESSEAMT. *Die Entwicklung des U-Bootkrieges seit Herbst 1917 im Urteil des Auslandes. . . .* Berlin, Hobbing, 1918. 128 pp. LC 25-15012.

Germany's High Sea Fleet in the World War. See entry 1097.

Gerry, B. See entry 778.

481 GERSON, Noel Bertram, 1914- . *Warhead.* NY, Doubleday, 1970. 336 pp. LC 78-103747.

A novel about the men who make nuclear submarines, their town and their women.

Die Geschichte der Deutschen Schiffsartillerie. See entry 1106.

Geschichte des Deutschen Ubootbaus. See entry 1053a.

Geschichte des U-Bootkrieges 1939-45. See entry 955.

482 GIBBS, Archie, 1906- . *U-Boat Prisoner: The Life Story of a Texas Sailor,* edited by Eugene Leuchtman. NY, Houghton, 1943. 208 pp. LC 43-11695.

483 GIBSON, John Frederic. *Dark Seas Above.* London, Blackwood, 1947. 286 pp. LC 48-20156.

484 GIBSON, Richard Henry, 1898- . *The German Submarine War, 1914-1918.* by R. H. Gibson and Maurice Prendergast. Foreword by ADM. Jellicoe. NY, Smith, 1931. xix, 438 pp., plates, ports, maps, diagrams, bibliog. LC 31-15312.

An essential book. The most complete and comprehensive history of the WW I U-boat operations available in English.

485 ________. *Germanskaia Podvodnaia Voina 1914-1918 gg.* Moscow, 1938. 431 pp., diagrams. LC 48-35996.

486 GIESSLER, Helmuth. *Der Marine Nachrichten und Ortungsdienst.* Munich, Lehmann, 1971. 156 pp., illus.

The German wireless and detection service, and ASW during WW II.

487 GILMORE, Horace Herman, 1903- . *Model Submarines for Beginners.* NY, Harper, 1962. 122 pp., with 87 pictures and diagrams by the author. LC 62-7944.

488 GILSON, Charles James Louis. *Submarine U-93: A Tale of the Great War.* London, Boys' Own Paper, 1916. 295 pp., illus.

489 GIMPEL, Herbert J. *The United States Nuclear Navy.* NY, Watts, 1965. 199 pp., illus., glossary. LC 65-21635.

Gimpel, Herbert Julius, 1915- . See entry 1175.

Giorli, Ezio. See entry 491.

Gladisch, Walter. See entry 535.

490 GLEAVES, Albert. *History of the Transport Service.* NY, Doran, 1921.

Gleeson, James Joseph. See entry 1422.

491 GIORLI, Ezio, 1864- . *Il Battello Sottomarino.* Roma, Tipografia Sociale, 1893. 24 pp. LC 17-14883.

Gli Angeli Senza Ali. See entry 272.

492 GLODSCHEY, Erich. *U-Boote, Deutschlands Scharfe Waffe.* Stuttgart, Union Deutsche Verlagsgesellschaft, 1943. 239 pp., illus. ports. LC A 50-4938.

493 GLUTH, Oskar Raoul, 1887- . *Wilhelm Bauer, der Er-*

finder des Unabhangigen Unterseeboots. Sein Werk und Sein Enttauschungen im Rahmen Seines Lebens Dargestellt. Munich, Hans Sachs, 1911. 58 pp., illus., ports.

Goat Power. See entry 437.

Goff, Loyal G. See entry 1210.

Golconda. See entries 521, 523.

The Golden Horseshoe. See entries 1041, 1042, 1045, 1046, 1048.

494 GOLOVKO, Arseni Grigorevich, 1906-1962. *With the Red Fleet: The War Memoirs of the late Admiral Arseni G. Golovko.* Translated by Peter Broomfield. Introduction by Sir Aubrey Mansergh. London, Putnam, 1965. 247 pp., illus., maps. LC 66-5065.

Good Hunting. See entry 192.

Goodman, M. W. See entry 958.

495 GORDEEV, I. *Krasny Morskoi Flot.* (The Red Navy). Moscow, 1925.

496 GORDON, Bernard L., ed. *Man and the Sea: Classic Accounts of Marine Explorations.* Foreword by Paul M. Fye. (Published for the American Museum of Natural History) NY, Doubleday, 1970. xxiv, 498 pp., illus. LC 76-116208.

Underwater exploration.

497 GORES, Joseph N., 1931- . *Marine Salvage: The Unforgiving Business of No Cure, No Pay.* Preface by Willard Bascom. NY, Doubleday, 1971. xxii, 525 pp., illus. LC 70-139025.

498 GORSHKOV, Sergei G. *Red Star Rising at Sea.* Annapolis, Naval Institute Press, 1974.

Originally appeared in the U.S. Naval Institute *Proceedings* as a translated series "Navies in War and Peace" by Gorshkov, Admiral of the Fleet of the Soviet Union. The articles included commentaries by American naval authorities. The book form includes an introduction by ADM. Elmo R. Zumwalt, Jr.

499 GORSKY, Bernard. *Vastness of the Sea: Adventure in the Mysterious Depths.* Translated by Alec Brown. Boston, Little, Brown, 1957. 305 pp., illus. LC 57-5829.

500 ________. *Moana (Vastness of the Waters).* Trans-

lated by Alec Brown. London, Elek Books, 1956. 239 pp., illus. LC 56-46412.

501 GOSSE, Joseph. *Les Mines Sous-Marines.* Paris, Pichon et Durand-Auzias, 1914. 185 pp., bibliography. LC 17-13640.

University of Paris thesis.

502 GOTTBERG, Otto von. *Kreuzerfahrten und U-Boots Taten.* Berlin-Wien, Ullstein, 1916. 215 pp.

503 GRAF, H. *The Russian Navy in War and Revolution from 1914 up to 1918.* Munich, 1923. NY, Kolands, 1923. vii, 233 pp., illus.

G. N. Kolands translated this from Russian into English.

Graph, HMS See entry 342.

504 GRAMLING, Homer. *Skin and Scuba Diving: A Basic Guide.* NY, Ronald, 1964. 184 pp., illus. LC 64-18464.

505 GRANT, Robert McQueen, 1917- . *U-Boats Destroyed: The Effect of Anti-Submarine Warfare, 1914-1918.* NY, Putnam, 1964. Chester Springs, Dufour, 1967. 172 pp., plates, maps, including 32 pp. of photos.

506 ________. *U-Boat Intelligence, 1914-1918.* NY, Putnam, 1969. 192 pp., illus., plates, maps. SBN 370-00034-X. LC 74-423831. Reprinted Hamden, Archon Books, 1969. LC 73-8392.

The Gravest 366 Days. See entry 383.

507 GRAY, Edwyn. A. *A Damned Un-English Weapon.* London, Seeley, Service, 1971.

Re the British submarine service during WW I.

508 ________. *The Killing Time. The U-Boat War 1914-18.* NY, Scribners, 1972. 241 pp., illus., bibliog., appendices.

509 ________. *The Underwater War: Submarines, 1914-1918.* NY, Scribners, 1972. 259 pp., illus., maps, bibliog. LC 76-178234.

Also issued as a paperback in 1973.

Grayfin, USS (fictional). See entry 1129.

510 GREAT BRITAIN. ADMIRALTY. *His Majesty's Sub-*

marines. London, HMSO, 1945. 64 pp., plates. LC 45-8799.

511 ________. *La Chasse aux Mines.* London, HMSO, 1943. 62 pp., illus. LC 44-2154.

512 ________. *German, Italian, Japanese U-Boat Casualties.* London, HMSO, 1946. 35 pp., tables.

513 ________. *Merchant Shipping losses . . . Return showing separately, British Merchant and Fishing Vessels Captured or Destroyed by the Enemy . . . 4th day of August 1914 to 11th day of November 1918. . . .* London, HMSO, 1919. xxii, 164 pp., tables. LC 20-8106.

514 ________. *Merchant Tonnage and the Submarine.* London, HMSO, 1918. 8 pp., tables, diagrams LC 18-11231.

Covers period August 1914-to December 1917.

515 ________. *Merchant Tonnage and the Submarine. Supplementary Statement . . . August 1914 to October 1918.* London, HMSO, 1918. 4 pp., tables, diagrams. LC 20-18112.

516 GREAT BRITAIN. ADMIRALTY. COMMITTEE ON DEEP-WATER DIVING. *Report of a Committee . . . Upon the Conditions of Deep-Water Diving.* London, HMSO, 1907. 79 pp., plates, diagrams. LC 8-14774.

Captain Frederick Tower Hamilton, 1856- , was President of the committee.

517 GREAT BRITAIN. CENTRAL OFFICE OF INFORMATION. *The Battle of the Atlantic.* London, HMSO, 1946. 103 pp. LC 47-304 rev.

518 GREAT BRITAIN. FOREIGN OFFICE. *Circular Dispatch Addressed to His Majesty's Diplomatic Representatives in Allied and Neutral Countries Respecting the Torpedoing by German submarines of the British Hospital Ships Rewa, Glenart Castle, Guildford Castle, and Llandovery Castle.* London, HMSO, 1918. 20 pp. LC 21-2055.

519 ________. *Correspondence with the German Government regarding the alleged misuse of British Hospital Ships.* London, HMSO, 1917. 26 pp. LC 17-31420.

520 ________. *Correspondence with the Swedish Govern-*

ment respecting the Mining of the Kogrund Passage. London, HMSO, 1917. 7 pp. LC 22-1412.

521 ________. *Correspondence with the United States Ambassador Respecting the Safety of Alien Enemies Repatriated from India on the SS Golconda.* London, HMSO, 1916. 4 pp. LC 16-25570.

522 ________. *Further Correspondence with the German Government Respecting the Incidents Alleged to have Attended the Sinking of a German Submarine and its Crew by H.M. auxiliary cruiser Baralong on August 19, 1915.* London, HMSO, 1916. 5 pp. LC 16-14964.

523 ________. *Further Correspondence with the United States Ambassador Respecting the Safety of Alien Enemies Repatriated from India on the S.S. Golconda.* London, HMSO, 1916. 3 pp. LC 17-24280.

524 ________. *Memorandum Respecting the Treatment of Belligerent Submarines in Neutral Waters. . . .* London, HMSO, 1916. 2 pp. LC 17-9506.

The Great Stone of Sardis. See entry 1187.

Great Undersea Adventures. See entry 1483.

The Greatest Depths. See entry 1152.

525 GREEN, John B., 1826- . *Diving With and Without Armor.* Buffalo, Faxon's Steam Power Press, 1859. vii, 62 pp. LC 18-4202.

Greenbaum, Leon Jack, Jr. See entries 616, 1211.

526 GREGOR, Rene. *The Russian Fleet in the First World War, 1914-17.* Edited by Ian Allen. Liverpool, 1972.

527 GRIDER, George W., 1912- . *Warfish* as told to Lydel Sims. Boston, Little, Brown, 1958. Toronto, Cassell, 1958. 282 pp. LC 58-7863.

Also a paperback ed. of 192 pp. pub. by Pyramid in 1961.
Story of the USS Flasher (SS 249) during WW II.

528 ________. *War Fish.* NY, Ballantine, 1973. 192 pp. Reissue of the 1958 ed.

529 GRIFFITH, George. *The Stolen Submarine.* London, White, 1904. 320 pp.

530 GRISHINSKY, A. S. and V. P. Nikolsky, eds. *Istoriia Russkoi Armii i Flota.* (A History of the Russian Army and Navy). Moscow, 1911- . 15 volumes.

531 GROELING, Dorothy Trautwein, 1908- . *Submarines, Disarmament and Modern Warfare.* Thesis. NY, Columbia University, 1935. vii, 199 leaves. Typed manuscript. Reissued in microfilm at Ann Arbor, by University Microfilms, 1950. Publication #1854. LC Mic A 50-121.

532 GROENER, Erich. *Die Deutschen Kriegsschiffe 1815-1936.* Munich, Lehmanns, 1937.

533 ________. *Die Deutschen Kriegsschiffe 1815-1945.* Munich, Lehmanns, 1966. 445 pp., illus., diagrams.

Data list of all the German warships which were built between 1815 and 1945.

534 ________. *Die Schiffe der Deutschen Kriegsmarine und Luftwaffe,* 1939-1945. 86 pp., 228 illus.

Grondige Oplossinge. See entry 336.

535 GROOS, Oto, 1882- and GLADISCH, Walter. *Der Krieg in der Nordsee.* Berlin, Mittler, 1920-1937. 7 volumes.

536 GROSSE, Karl Freidrich, 1913- . *Politische und Militärische Bedeutung des Unterseebootskrieges 1914/18.* Rosenheim, Druck Rosenheimer, 1937. 163 pp., diagrams. LC 40-17089.

537 GROSSETT, Harry. *Down to the Ships in the Sea.* Philadelphia, Lippincott, 1954. 256 pp., illus. LC 54-9423.

Submarine rescue and salvage, especially re HMS Affray.

Grosvenor, Gilbert M. See entry 347.

538 GROSVENOR, Joan. *Open the Ports: The Story of Human Minesweepers.* with L. M. Bates. London, Kimber, 1956. 199 pp., illus. LC 56-42838.

WW II underwater demolition teams.

539 GROTIUS SOCIETY, LONDON. *Discussion on the Abolition of Submarines.* (In Grotius Society—*Problems of Peace and War.* London, 1926. Vol. 11, p. 65-78.) LC A 26-394.

540 ________. Report of the Committee on the Legal Status of Submarines. (In Grotius Society. *Problems of Peace*

and War. London, 1929. vol. 14, p. 155-174) LC A 30-471.

Groups Under Stress. See entry 1009.

Gründliche Aufflosung. . . . See entry 337.

541 GRUSS, Robert. *The Art of the Aqualung*. Translated from "Manuel de l'homme sans poids" by Richard Garnett. London, MacGibbon & Kee, 1955. 66 pp., illus. LC 56-15515.

Guerilla Submarines. See entry 314.

Guerilla Warfare. See entry 661.

La Guerra al Commercio con Sommergibili. See entry 1166.

La Guerra Submarina de Alemania. See entry 449.

La Guerra Sui Mari. See entry 103.

La Guerre de Mines. See entry 141.

La Guerre Sous les Mers. See entry 298.

La Guerre Sous-Marine. See entry 1167.

La Guerre Sous-Marine et les Torpedos. See entry 286.

La Guerre sur Mer, Stratégie et Tactique. See entry 285a.

542 GUGLIELMOTTI, Leandro. *La Nave Sommergibili, Come Construita, Come Funziona, Come Opera*. Preface by ADM Guido Milanesi. Milan, Society Anomina Editrice Dante Alighieri, 1931. xiv, 297 pp., illus., diagrams, ports.

Guiá Submarina de la Costa Brava. See entry 1031.

Guide for the Classification of Manned Submersibles. See entry 14.

Guide to the Soviet Navy. See entry 145.

Guide to the Underwater. See entry 1142.

543 GUIERRE, Maurice Casimir Lucien, 1888- . *L'Angoisse des Veilles Sousmarines*. Preface by André Saures. Paris, Floury, 1919. iv, 22 pp., illus. LC 19-16126.

A poem, privately printed in a limited edition of 800 copies, with woodcut illustrations by the author.

544 ————. *L'Epopee du Surcouf et le Commandant Louis Blaison*. Paris, Bellenand, 1952. 250 pp., illus.

545 ________. *Sous-Mariniers.* Paris, Flammarion, 1948. 306 p. LC 49-27005.

546 GUIHENEUC, Olivier Francois Henri. *Dreadnought ou Submersible?* Paris, Perrin, 1916. 316 pp., plates.

547 GUILLERME, Jacques, 1927- . *Traité de Plongée* by Jacques Guillerme and Jean Rivoire. Preface by Pierre Drach. Paris, Dunod, 1955. 213 pp., illus. LC 55-29998.

548 GUITARD, Pierre. *U-Boat Warfare.* London, Pallas, 1939. 80 pp. LC 40-32452.

Gunnery, Naval. See entry 1106.

549 GWATKIN-WILLIAMS, Rupert Stanley. *Under the Black Ensign.* London, Hutchinson, 1922. 238 pp., illus., map. LC 22-15653.

Activities of HMS Intrepid in the Russian Arctic during WW I.

H

H.3 (submarine). See entries 859, 860.

550 HAAS, Walter de, 1886- . *Versunkene Schätze; Die Eroberung der Tiefe.* Stuttgart, Franckh, 1928. 79 pp., illus. LC 31-12503.

Illustrated by R. Oeffinger.

551 HACKFORTH-JONES, Gilbert, 1900- . *No Less Renowned.* Edinburgh and London, Blackwood, 1939. 306 pp. LC 39-16503.

552 ________. *Submarine Alone: A Story of HMS Steadfast* with illustrations by Douglas Wales. London, Hodder, 1943. Toronto, Musson, 1943. 141 pp., illus. LC 44-1738.

553 ________. *Submarine Flotilla. A Chapter in the Life of an Obedient Servant.* London, Hodder, 1940. Toronto, Musson, 1940. 283 pp. LC 40-8820.

554 ———. *True Book About Submarines.* Illustrated by David Cobb. London, Muller, 1955. Sydney, Shakespeare, 1956. 141 pp., illus.

Hadfield, Robert L. See entry 1456.

555 HAGUE. INTERNATIONAL PEACE CONFERENCE. 2nd, 1907. *Convention between the United States and Other Powers Relative to the Laying of Automatic Submarine Contact Mines.* Washington, GPO, 1910. 23 pp. (Treaty Series, no. 541) LC 16-15969.

556 ———. *The Hague Convention (VIII) of 1907 Relative to the Laying of Automatic Submarine Contact Mines.* Washington, Carnegie Endowment for International Peace, 1915. 6 pp. LC 15-5926.

557 HAHN, Walter A., 1921- . *On Using Self Contained Underwater Breathing Apparatus,* with Christian J. Lambertsen. Washington, 1952. 19 pp. (National Academy of Sciences. National Research Council. Pub. 274) LC 54-4855.

Hale, William Bayard, 1869-1924. See entry 1284.

Half Mile Down. See entry 82.

558 HALL, Cyril, 1884- . *Modern Weapons of War, by Land, Sea, and Air.* London, Blackie, 1915. 192 pp., illus. LC War 15-9.

559 HALL, Sydney Stewart, 1872- . *Submarine Warfare.* (In Grotius Society, London. *Problems of Peace and War.* London, 1920. Vol. V, p. 82-94) LC A 21-1050 Revised.

Hall, ADM. Sydney Stewart, 1872- . See entries 190, 191.

Halley, Edmond. See entry 1419.

560 HAMILTON, Lee David. *Let's Go Aboard an Atomic Submarine.* NY, Putnam, 1965. Don Mills, Longmans, 1965. 48 pp., illus. by Robert Bartram. LC 65-10876.
Juvenile.

Hamm, C. B. See entry 1223.

Hammond, John Hays, Jr. See entry 417.

561 HAMMOND, Rolt. *Diving and Salvage.* London, Muller, 1964. New Rochelle, Soccer, 1964. 143 pp., illus. LC 64-6253.

Hampton Roads, Va. See entry 271.

Handbook for Skin Divers. See entries 152,278.

Handels-U-Boot Deutschland. See entry 734.

Handels-U-Boot Deutschland Fahrt Nach Amerike. See entry 750.

Der Handelskrieg mit U-Booten. See entry 1165.

562 HARDEGEN, Reinhard. *Auf Gefechtsstationen!* Leipzig, Boreas, 1943. 227 pp., illus.

"The flamboyant skipper of U-123 gives his version of the U-boat war in American waters in 1942-1943."—Farago.

Harder, USS (SS 257). See entry 793.

563 HARDY, René. *The Lost Sentinel.* Paris, Robert Laffont, 1959. Garden City, Doubleday, 1960. (Humphrey Hare, tr.) 223 pp. LC 60-12437. London, Laurie, 1960. LC 60-29902.

A novel about the Italian submarine Moro in the Mediterranean during World War Two.

564 HARDY, William Marion. *Submarine Wolfpack.* NY, Dodd, Mead, 1961. 175 pp. LC 61-12917.

565 ________. *A Time of Killing.* NY, Dodd, Mead, 1962. London, Hamilton, 1962. 186 pp. LC 62-9719.
9719.

A novel about American submarines in the Pacific by a WW II submariner.

566 ________. *A Time of Killing.* NY, Dell, 1963. 192 pp.

567 ________. *U.S.S. Mudskipper. A Novel of WW II.* NY, Dodd, Mead, 1967. 224 pp.

568 ________. *Wolfpack.* NY, Dodd, Mead, 1960. 183 pp. LC 60-8439.

Harrington, Joseph D. See entries 1486, 1487.

569 HARRIS, L. E. *The Two Netherlanders: Humphrey Bradley and Cornelis Drebbel.* Leiden, Brill, 1961. viii, 228 pp., illus.

570 HARRIS-WARREN, Herbert Benjamin, 1921- . *Dive! The Story of an Atomic Submarine.* NY, Harper, 1960. 130

pp., illus. W 60-5778.

For grades 5 through 9.

571 HART, Sydney. *Discharged Dead: A True Story of Britain's Submarines at War.* London, Oldhams, 1956. 208 pp., plates.

572 ________. *Submarine Upholder.* London, Oldbourne Press and Daily Express, 1960. 208 pp.

573 HARTMANN, Werner. *Feind im Fadenkreuz. U-Boot auf Jagd im Atlantic.* Nacherzalt von Gerhart Weise, mit einem vorwort vom befehlshaber der U-Boot Vizeadmiral Karl Doenitz. Berlin, Heimbucherei, 1942. 232 pp., illus., plates.

WW II German submarine operations in the Atlantic.

574 HASHAGEN, Ernst. 1885- . *The Log of a U-Boat Commander; or, U-Boats Westward—1914-1918.* London and NY, Putnam, 1931. 277 pp., frontis., plates, ports, maps. Translated by Vesey Ross. LC 31-34793.

575 ________. *U-Boats Westward!* NY, London, Putnam, 1931. xiii, 247 pp., frontis., plates, ports, maps. Translated by Vesey Ross. LC 31-28431.

576 ________. *U-Boote Westwärts! Meine Fahrten um England, 1914-1918.* Berlin, Mittler, 1931. xi, 219 pp., illus., plates, ports, maps. LC 31-15805.

577 ________. *U-Boote Westwärts! Meine Fahrten um England, 1914-1918.* New edition. Berlin, Mittler, 1942. 158 pp., illus., plates, ports, map. LC AF 48-487.

578 HASHIMOTO, Mochitsura. *Sunk: The Story of the Japanese Submarine Fleet, 1941-1945.* Translated by E. H. M. Colegrove. Introduction by Edward L. Beach. NY, Holt, 1954. 276 pp., illus.

Includes account of the sinking of USS Indianapolis by the I-58.

579 ________. *Sunk! The Story of the Japanese Submarine Fleet, 1941-1945.* NY, Avon, 1954. (T 246) 190 pp., appendices.

Paperback abridgement.

580 ________. *Les Sous-Marins du Soleil Levant.* Paris, Presses de la Cite, 1955.

581 HASS, Hans. *Ich Fotografierte in den 7 Meeren.* See-

bruck am Chiemsee, Heering, 1955 (c 1954). 160 pp., illus. LC 56-34402.

Underwater photography.

Hastings, Priscilla. See entry 1056.

582 HAUFF, Louis. *Die Unterseeische Schiffahrt.* Munich, 1859. Bamberg, Buchners, 1915. 78 pp., illus.

About Wilhelm Bauer, 1822-1876, and his work with submarines.

Havre. See entry 210.

583 HAWKINS, Doris Maxwell. *Atlantic Torpedo.* London, Gollancz, 1943. 48 pp. LC 44-2621.

584 HAWKINS, Maxwell. *Torpedoes Away, Sir! Our Submarine Navy in the Pacific.* NY, Holt, 1946. Toronto, Oxford, 1946. 268 pp., illus. LC 47-30018.

585 HAY, Marley Fotheringham. *Secrets of the Submarine.* NY, Dodd, Mead, 1917. ix, 229 pp., frontis, plates. LC 17-24865.

586 HAYDON, A. L., ed. *Model Yachts, Sailing Boats and Submarines.* London, Boys' Own Paper, n.d. (ca. 1910). 190 pp., illus.

Pp. 40-110 has section on submarine models by V. E. Johnson and H. F. Hobden.

587 HAYES, Clair Wallace, 1887- . Wrote the *Boy Allies With the Navy* series under the pseudonym of Robert L. Drake. There were ten volumes in the series all published in NY by A. L. Burt from 1915-22. They sold at 60¢ per volume. Three of the titles dealt with submarines, *viz:*

The Boy Allies with the Terror of the Sea: or, The Last Shot of Submarine D-16.

The Boy Allies Under the Sea: or, The Vanishing Submarine.

The Boy Allies With the Submarine D-32; or, The Fall of the Russian Empire.

588 HAYES, Hammond Vinton, 1860- . *Submarine Signaling, Its Application in Peace and War.* Boston, The Submarine Signal Co., 1920. 27 pp., frontis, illus., maps. LC 22-2506.

589 HAZELTINE, Alice Isabel, comp. *Below the Surface.* Nashville, Abingdon Press, 1958. 223 pp.

590 HAZLETT, Edward Everett, 1892- . *Rig For Depth Charges! The Career of a Young Naval Officer on Submarine Duty.* NY, Dodd, Mead, 1945. 269 pp. LC 45-8598.

591 HEIMBURG, Heino von. *U-Boot Gegen U-Boot.* Berlin, Scherl, 1917. 124 pp. LC 42-46792.

Die Helden der U-Boote. See entry 126.

Hell at 50 Fathoms. See entry 789.

Hell Below. See entry 369.

Hellcats of the Sea. See entry 790.

Hell's Angels of the Deep. See entry 193.

Helmreich, Robert. See entry 1009.

Hendrick, Burton Jesse, 1870- . See entries 1136, 1177.

Henley, M. W. See entry 145.

Henriques, H. See entry 873.

592 HERBERT, Frank. *The Dragon in the Sea.* Garden City, Doubleday, 1956. 192 pp. LC 56-5586.

593 ________. *Twenty-First Century Submarine.* NY, Avon, 1956.

Paperback version of *The Dragon in the Sea.*

Herbert, Godfrey, 1884- . See entry 213.

594 HERLIN, Hans. *Verdammter Atlantik.* Hamburg, Nannen, 1959. 426 pp., illus.

"A dramatized narrative of the highlights of the U-Boat war by a German journalist. Contains much material not otherwise available."—Ladislas Farago.

595 HERRICK, John Origen, 1898- . *Subsurface Warfare: The History of Division 6, NDRC.* Washington, Dept. of Defense, Research and Development Board, 1951. 135 pp., ports. LC 51-60645.

596 HERRICK, Robert Waring. *Soviet Naval Strategy. Fifty Years of Theory and Practice.* Annapolis, U.S. Naval Institute, 1968. xxxiv, 197 pp. LC 67-26080.

Herrington, Roger. See entry 188.

597 HERSEY, John Brackett, ed. *Deep-Sea Photography.* (The

Johns Hopkins Oceanographic Studies, No. 3) Baltimore, Johns Hopkins Press, 1967. 352 pp., illus. LC 66-16038.

25 original papers by practicing experts in underwater photography.

598 HERSING, Otto. *U-21 Rettet die Dardanellen.* Leipzig, Hase & Koehler, 1932. 151 pp., illus., port. LC 51-52267.

Hervet, R. See entry 1076.

599 HERZOG, Bodo. *Claus Bergen: Leben und Werk des Grossen Marinemalers.* Krefeld, Ruhl, 1963. 117 pp., illus., port.

Bergen's paintings are hanging at USN Submarine Base at New London and at Norfolk.

600 ________. *Die Deutschen U-Boote: 1906-1945.* Munich, Lehmanns, 1959. 234 pp., illus., tables, folding chart.

An important data book.

601 ________. *U-35: Das Erfolgreichiste Unterseeboot der Welt! Kapitänleutnant Lothar von Arnauld de la Periére, der Erfolgreichste U-Boot-Kommandant der Welt.* Foreword by Charles A. Lockwood. Krefeld, Ruhl, 1964.

602 ________. *Kapitänleutnant Otto Steinbrinck. Die Geschichte des Erfolgreichsten U-Boot-Kommandanten in den Gewassern um England.* Krefeld, Ruhl, 1963. 233 pp., illus., ports, bibliog.

603 ________. *60 Jahre Deutsche U-Boote. 1906-1966.* Munich, Lehmanns, 1968. 324 pp., illus. Schematics.

Covers pre-WW I, WW I, WW II, and German Federal Forces U-Boats. Excellent source of historical and technical facts.

604 ________. *U-Boote im Einsatz 1939-45.* Dorheim, Podzun, 1971. 256 pp., illus.

"U-Boats in Action." Pictorial documentary with text in German and in English.

605 HERZOG, Bodo and Günter SCHOMAKER. *Ritter der Tiefe-Graue Woelfe.* (Knights of the Deep-Gray Wolves). Munich, Welsermuehl, 1965. 563 pp., illus.

Biographies of the most successful U-Boat and submarine commanders of WW I and WW II, both friend and foe.

He's In Submarines Now. See entry 397.

He's In the Sub-Busters Now. See entry 1013.

Hess, Lowell. See entry 1184.

606 HEWLETT, Richard G. and Francis DUNCAN. *Nuclear Navy, 1946-1962.* 1974. 544 pp., illus., charts.

"Beginning with the germ of the idea in 1939, this illustrated survey traces the growth of the U.S. Navy's nuclear fleet through 1962, when 27 submarines and three surface ships were in operation." —USNI brochure.

607 HEZLET, Sir Arthur Richard, 1914- . *The Submarine and Sea Power.* London, Davies, 1967. Plymouth, Clark, Doble & Brendon, 1967. LC 67-85112. NY, Stein and Day, 1967. (LC 67-15760.) 278 pp., illus., plates, maps.

608 HIGGINS, Alexander Pearce, 1865- . *Defensively-Armed Merchant Ships and Submarine Warfare.* (Abridged ed.) London, Stevens, 1917. 32 pp. LC 38-23426.

609 HIGGINS, Edward T. *Webfooted Warriors: The Story of a Frogman in the Navy During World War II.* In collaboration with Dean Phillips. NY, Exposition Press, 1955. 172 pp. LC 54-13405.

Hildebrand, H. H. See entry 795.

610 HILDESCHEIM, Erik. *Med Staalnerver: Motorfolkene i Verdenskrigen paa Jorden, Vandet og i Luften.* Copenhagen, Lehmann & Stage, 1915. 171 pp., illus. LC 21-14761. 14761.

611 HINKAMP, Clarence Nelson. *Submarines and Torpedoes.* Abstract of an address delivered before the (Phi Beta Kappa) Association on November 20, 1914. Washington, 1914. 2 pp. LC CA 18-155 Unrev'd.

Hinsdale, William Russell. See entry 1455.

His Majesty's Submarines. See entry 510.

H.M.U-Boat. See entry 342.

H.M.Submarines. See entry 699.

H.M.S.Thule Intercepts. See entry 834.

Histoire de la Marine Russe. See entry 874.

Histoire du Torpilleur en France 1872-1940. See entry 765.

Histoire Mondiale du Sous-Marin. See entry 26.

Histoire Sous-Marine des Hommes. See entry 412.

Historic Submarines. See entry 785.

History and Development of Submarine Signals. See entry 393.

A History of Russian and Soviet Sea Power. See entry 868.

A History of Seafaring Based on Underwater Archaeology. See entry 60.

A History of Self-Contained Diving and Underwater Swimming. See entry 749.

History of Submarine Mining and Torpedoes. See entry 360.

A History of the French Navy. See entry 685.

A History of the Russian Army and Navy. See entry 530.

History of the Transport Service. See entry 490.

History of Torpedo Warfare. See entry 137.

History Under the Sea. See entry 820.

History Under the Sea. See entries 966, 967.

612 HOAR, Allen. *The Submarine Torpedo Boat, Its Characteristics and Modern Development.* NY, Van Nostrand, 1916. xv, 211 pp., illus., folding plates, diagrams. LC 16-18778.

613 HOBAN, Russell. *The Atomic Submarine: A Practice Combat Patrol Under the Sea.* NY, Harper, 1960. London, Hamilton, 1960. folio. Unpaged. Illus. LC 60-9454.
For grades 2-5.

614 HOCHSTETTER, Franz, 1880- . *Ratsel der U-Bootswirkung.* Berlin, Politik, 1919. 31 pp. LC 21-4086.

615 HOEHLING, Adolph A. and Mary. *The Last Voyage of the Lusitania.* NY, Holt, 1956. Toronto, McLeod, 1956. 255 pp., plates. LC 56-7724.

Die Hollenmaschine im U-Boot. See entry 1087.

616 HOFF, Ebbe Curtis and Leon Jack GREENBAUM, Jr. *A Bibliographical Sourcebook of Compressed Air, Diving, and*

Submarine Medicine. LC 48-46250.

v. I. Through 1946. First published 1948, reprinted 1967. 382 pp., illus.

v. II. Covers the literature from 1/1/46 through 12/31/51. First published 1954. Reprinted 1967. 386 pp.

v. III. Covers the literature from 1/1/52 through 12/31/61, with some references through 1964. 1967. 306 pp.

Available from Superintendent of Documents. U.S. Government Printing Office. Washington, D.C. (D 206.2:M 46)3 v. 1,2,3)

Hogue, HMS See entry 1441.

617 HOLBROOK, Stewart Hall, 1893- . *Lost Men of American History.* NY, London, Macmillan, 1946. xiv, 370 pp., plates. LC 46-7135.

Includes story of Ezra Lee (1749-1821) and the Turtle.

618 ________. *None More Courageous: American War Heroes of Today.* NY, Macmillan, 1942. London and Toronto, Macmillan, 1943. 245 pp., plates. LC 42-51084.

Holland, John Philip, 1842-1914. See entries 175, 473, 885, 886, 1177.

619 HOLLAND (submarine). U.S. Navy Department. *Copies of all Reports in the Possession of the Navy Department and Extracts of Reports on File in the Bureau of Navigation Concerning the Submarine Boat Holland.* 56th Congress, 2nd Session. Senate Doc. 71. Washington, 1900. 13 pp.

Holland (submarine). See also entry 619.

Holland submarines. See entry 746.

Holland Torpedo-Boat Company. See entry 1288.

620 HOLLWEG, Karl. *Unser Recht auf den U-Bootskrieg.* Berlin, Ullstein, 1917. 242 pp. LC 23-18486.

621 HOLMES, Wilfred Jay, 1900- . *Undersea Victory: The Influence of Submarine Operations on the War in the Pacific.* NY, Doubleday, 1966. 528 pp., charts, maps, bibliog. LC 65-17232.

Discusses the submarine efforts of all the nations involved in the Pacific during WW II.

Holmes, Wilfred Jay, 1900- . See also Alec Hudson, pseud.

Holmquist, June Drenning. See entry 245.

L'Homme-Poisson et se Conquetes. See entry 1064.

Los Hombres-Peces. See entry 1032.

622 HONOUR, Alan, 1918- . *The Adventures of the Piccards.* London, Brockhampton, 1959. 128 pp., plates.

623 ————. *Ten Miles High, Two Miles Deep: The Adventures of the Piccards.* NY, McGraw-Hill, 1957. Toronto, 1957. 206 pp., illus. LC 57-9423.

624 HOOGENDIJK, Johannes Hendrikus, 1889, comp. *De Nederlandsche Koopvaardij in den Oorlogstijd (1914-1918) Eigen Ervaringen van Gezagvoerders. . . .* Amsterdam, Holkema & Warendorf, 1930. 508 pp., illus., maps, facsims., plates. LC 33-10789.

Personal narratives of the experiences of Dutch merchant seamen during WW I.

Hopkins, Gerard. See entry 309.

625 HOPKINS, John Jay. *We Work For Freedom.* NY, Newcomen Society, 1950. 28 pp. LC 51-2499.

Transcript of an address by the president of the Electric Boat Co.

Hopkins, Sir John Ommaney, 1834- . See entry 328.

Hopkins, Nevil Monroe, 1873- . See entries 147, 148.

626 HOPKINS, William John. *The Clammer and the Submarine.* Boston, Houghton Mifflin, 1917. 346 pp.

Fiction re German submarine menace off East coast of USA.

Horan, James David, 1914- . See entry 429.

Horrabin, J. F. See entry 869.

627 HORTON, Edward. *The Illustrated History of the Submarine.* London, Sidgwick and Jackson, 1974. 160 pp., illus.

Horton, Sir Max Kennedy. See entry 207.

628 HOSMER, Helen R., comp. *Submarines in Periodical Literature, 1911-1917.* Phila., The Franklin Institute, 1917.

Reprint from their *Journal* v. 184, August 1917. pp. 251-306.
A list of selected articles with annotations.

629 HOUGH, Horatio Gates. *Diving or an Attempt to Describe Upon Hydraulic and Hydrostatic Principles, A Method of Supplying the Diver With Air Underwater.* Hartford, Russell, 1813. 7 pp., diagram. LC 6-5668.

630 HOULDER BROS. & CO. LTD. *Sea Hazard (1939-45).* London, 1946.

631 HOUOT, Georges, 1913- and P. H. WILLM. *2000 Fathoms Down.* Introd. by Jacques-Yves Cousteau. Preface by Philippe Tailliez. Translated by Michael Bullock. NY, Dutton, 1955. 192 pp., illus., plates. (LC 55-8323) London, Hamilton, 1955. 256 pp., plates. (LC 55-3045).

632 HOVGAARD, William, 1857- . *Submarine Boats.* London & NY, Spon, 1887. 98 pp., illus., folding plates. LC 16-25824.

How Atomic Submarines are Made. See entries 248, 249.

Howard, William E. See entry 38.

633 HOYT, Edwin Palmer, 1923- . *From the Turtle to the Nautilus: The Story of Submarines.* Illustrated by Charles Geer. Boston, Little, Brown, 1963. 134 pp., illus. LC 63-7117.

634 ________. *The Sea Wolves: Germany's Dreaded U-Boats of World War Two.* NY, Lancer Books, 1972. 160 pp., illus. LC 72-188996.

Cover title: *The True and Gripping Saga of the Sea Wolves: Germany's Dreaded U-Boats of WW II.*

635 HUAN, Claude. *L'Enigme des Sous-Marins Sovietiques.* Paris, France-Empire, 1959. 297 pp., illus. LC 60-32580.

636 HUBERNAGEL, Wilhelm, 1903- . *Die Bekanntmachungen des U-Boot-Handelskrieges vom 4 Februar und 1 Februar 1917.* Greifswald, Hartmann, 1927. 89 pp. LC 44-49734.

Submarines and international law.

637 HUDSON, Alec, pseud. *Battle Stations!* NY, Macmillan, 1940. Toronto, Macmillan, 1940. 71 pp. LC 40-27152.

638 ________. *Enemy Sighted.* NY and Toronto, Macmillan, 1941. 61 pp. LC 41-51572.

639 ________. *Open Fire!* NY and Toronto, Macmillan, 1942. LC 42-15691.

Collection of short stories includes: Battle Stations, Enemy Sighted, Rendezvous, and Night Action.

640 ________. *Rendezvous.* NY and Toronto, Macmillan, 1942. 94 pp. LC 42-36023.

641 ________. *Rendezvous, and Other Long and Short Stories About Our Navy in Action.* Foreword by Carl Van Doren. NY, Readers Club, 1943. vii, 248 pp. LC 43-17321.

642 ________. *Submarine.* London, Nicholson, 1943. 160 pp.

Hudson, Alec is the pseudonymn of Wilfred Jay Holmes (q.v.).

Hudson, Lynton. See entry 430.

Hughes, R. T. C. See entry 1425.

643 HUIDOBRO TORO, Alfonso G. *El Submarino Ante el Derecho Internacional, Ensayo: Memoria de Prueba Para Optar al Grado de Licenciado en la Facultad de Ciencias Juridicas y Sociales.* Santiago de Chile, Talleres Graficos Gutenberg, 1936. 185 pp., bibliog. LC 44-44342.

Submarine and international law. Apparently a thesis.

Hunley, C.S.S. See entry 352.

Hunley, Horace D. See entries 51, and 352.

Hunt, J. L. F. See entry 215.

Hunter, Anthony. See entry 1115.

The Hunters and the Hunted. See entry 144.

The Hunters and the Hunted. See entries 230, 231.

Hunting the German Shark. See entry 1450.

Hural, Jack. See entry 665.

644 HURD, Archibald Spicer, 1869- . *Italian Sea-Power and the Great War.* London, Constable, 1918. 124 pp.

pp. 79-93 deal with Italian submarines of WW I.

645 ________. *Murder at Sea.* London, Unwin, 1916. . . . 38 pp. LC 16-14717 Revised.

646 ________. *Ordeal by Sea: The Story of the British Seaman's Fight for Freedom.* London, Jarrolds, 1918? 227 pp., frontis. plate. LC 18-10562.

647 ________. *Outlawry at Sea: An Indictment of the German Navy.* London, Chapman and Hall, 1915. 15 pp. LC 19-13941.

Reprinted from *The Fortnightly Review* July, 1915.

648 ________. *Submarines and Zeppelins in Warfare and Outrage.* London, Causton, 1916. 22 pp. LC War 16-169 rev.

Hurd, Archibald Spicer, 1869- . See also Entry 41

Hurtado, Genaro. See entry 170.

Hush: or, The Hydrophone Service. See entry 1472.

Huston, John, d. 1968. See entry 694.

649 HUTCHISON, Miller Reese, 1876- . *The Submarine Boat Type of Edison Storage Battery.* Orange, N.J., 1915. 28 pp., illus.

650 HUTTER, Julien Eugene, 1874- . *Les Sous-Marins.* Paris, Berger-Levrault, 1917. 11 pp. (Fascicle 130 of *Pages d'Histoire, 1914-1917.*)

Hvorledes og Hvorfor med Nautilus. See entry 1206.

651 HYATT, A. M. J., ed. *Dreadnought to Polaris.* Annapolis, Naval Institute Prss, 1973. 125 pp., illus.

Papers presented at the 1972 Conference on Strategic Studies.

Hydrographie Contenant. . . . See entry 423.

Hydrospace. See entry 176.

I

I-58 (Japanese submarine). See entries 578, 579, 580.

I Dive for Treasure. See entry 1036.

I Like Diving. See entry 356.

I Sank the Royal Oak. See entry 998.

Ice Station Zebra. See entry 821.

652 ICENHOWER, Joseph Bryan, 1913- . *The First Book of Submarines.* Pictures by Mildred Waltrip. NY, Watts, 1957. 60 pp., illus. LC 56-5855.
Juvenile.

653 ________. *Submarines in Combat.* NY, Watts, 1964. Toronto, Ambassador, 1964. 180 pp., illus. LC-17786.
Juvenile.

654 ________. *Submarine Rendezvous: Pacific Adventure in World War II.* Illustrated by Norman Guthrie Rudolph. NY, Winston, 1957. 182 pp., illus. LC 57-5225.
Juvenile.

Ich Fotografierte in den 7 Meeren. See entry 581.

Ictineo (Spanish submarine). See entries 879, 880, 881.

I. D. New Tales of the Submarine War. See entry 845.

655 IDRIESS, Ion Llewellyn, 1890- . *Forty Fathoms Deep.* Sydney and London, Angus & Robertson, 1939. 343 pp., frontis., plates, ports. LC 40-30109.
Diving for pearls.

656 ________. *Forty Fathoms Deep: Pearl Divers and Sea Rovers in Australian Seas.* 5th ed. Sydney, Angus & Robertson, 1937. 343 pp., frontis., plates, ports. LC 38-2508.

657 IDYLL, Clarence Purvis, 1916- . ed. *Exploring the Ocean World: A History of Oceanography.* NY, Crowell, 1970. 280 pp., illus., maps. LC 78-78268.

658 IGNAT'EV, K. F. *Teoriia Podvodnykh Lodok.* Moscow, 1947. 186 pp., diagrams. LC 48-26083.

De Ijzeren Zeehond. See entry 1230.

The Illustrated History of the Submarine. See entry 627.

Im U-Boot Durch die Weltmeere. See entry 1107.

The Imperial and Royal Austro-Hungarian Navy. See entry 1147.

The Imperial Japanese Navy. See entry 674.

Imperial Russian Navy. See entries 675, 676.

659 IMPERIAL WAR MUSEUM. LONDON. *Bibliography of Submarine Warfare.* London, 1953. 16 pp. LC 57-43786.

660 ________. *Submarine Warfare World War I: Miscellaneous Articles.* London, 1957. 9 pp.

In Balloon and Bathyscaphe. See entry 970.

In Der Alarmkoje von U35. See entry 395.

Incentive pay. See entry 1295, 1296.

Indianapolis, U.S.S. See entries 578, 579, 580, 912.

Infernal Machines. See entry 963.

661 INGHAM, Travis, 1906- . *Rendezvous by Submarine: The Story of Charles Parsons and the Guerilla-Soldiers in the Philippines.* Foreword by Carlos P. Romulo. Garden City, Doubleday, 1945. Toronto, McClelland, 1945. 255 pp. LC 45-8622.

662 INSTITUTE ON UNDERWATER ACOUSTICS. LONDON. *Underwater Acoustics.* Proceedings of an institute sponsored by the Scientific Affairs Committee of the NATO and conducted by the Pennsylvania State University at the Imperial College of Science and Technology of the University of London. July 31-August 11, 1961. Edited by V. M. Albers. NY, Plenum Press, 1963. 354 pp., illus. LC 62-8011.

663 INTERAGENCY COMMITTEE ON OCEANOGRAPHY. *Undersea Vehicles for Oceanography.* Washington, ICO Pamphlet No. 18, 1965. 81 pp., illus., diagrams, bibliog. (Item 1061-B) LC 66-60407.

International Agreement for Collective Measures Against Piratical Attacks in the Mediterranean by Submarines. See entry 932.

International Law. See entries 66, 141, 159, 173, 205, 427, 461, 501, 524, 636, 643, 668, 883, 900, 984, 1022, 1050, 1051, 1148, 1168, 1194, 1263, and 1315.

Internationale Submarine Bibliographie. See entry 666.

Intrepid, H.M.S. See entry 549.

Introduction à L'Étude de la Guerre Sous-Marine. See entry 757.

Inventions and Devices. See entry 134.

The Invisible Destroyer. See entry 913.

Iron Coffins. See entry 1444.

Isaac Peral (Spanish submarine). See entry 824.

Isaac Peral; Le Tragedia del Submarino Peral. See entry 962.

664 ISAACS, Edouard Victor Michael. *Prisoner of the U-90.* Foreword by Josephus Daniels. Boston, Houghton, 1919. 185 pp., illus. LC 19-6417.

665 ISAKOV, Ivan Stepanovich. *The Red Fleet in the Second World War.* Translated by Jack Hural. London, Hutchinson, 1948. 124 pp., illus., maps. LC A48-3666.

666 I.S.B.: INTERNATIONALE SUBMARINE BIBLIOGRAPHIE. *Bibliography.* Braunschweig, Germany. 1954— (annual).

Title varies.

Issledovaniia i Izyskaniia. See entry 939.

Istoriia Russkoi Armii i Flota. See entry 530.

The Italian Navy in WW II. See entry 139.

Italian Sea-Power and the Great War. See entry 644.

Italian Submarines. See entries: 103, 129, 139, 142, 181, 230, 231, 272, 273, 329, 407, 512, 542, 563, 644, 667, 859, 860, 987, 1150, 1257.

Italy. See entries 1286, 1293, 1299, 1310, 1317, 1320.

667 ITALY. MARINA. STATO MAGGIORE. UFFICIO STORICO. *I Nostri Sommergibili Durante la Guerra 1915-18.* Rome, Tip. dell' Ufficio del Capo di Stato Maggiore, 1933. 331 pp., diagrams. LC 46-40529.

668 ITALY. PARLAMENTO. CAMERA DEI DEPUTATI. COMMISSIONE SUGLI RAPPORTI POLITICA CON L'ESTERO COLONIE. *Relazione della Commissione Rapporti Politici con L'Estero Colonie sul Disegno di Legge. . . .* Rome, 1922. 9 pp. LC 25-16450.

International law and the submarine and arms limitation talks at the Washington Conference on the Limitation of Armament, 1921-22.

J

669 JACKSON, George Gibbard, 1877- . *The Romance of the Submarine*. Phila., Lippincott, 1930. 244 pp., frontis., plates. LC 30-23413.

Jagd auf de Woelfe. See entry 1043.

Jakten på U-Båtar. See entry 928.

670 JAMES, Alan, 1943- . *Submarines*. Illustrated by John Beswick. Oxford, Blackwell, 1973. 61 pp., illus. ISBN 0-631-06940-2.

Juvenile.

671 JAMES, Henry Johnson, 1902- . *German Subs in Yankee Waters: First World War*. Illustrated by Charles E. Pont. NY, Gotham House, 1940. 208 pp., illus. LC 40-30476.

Jameson, Mary Ethel. See entry 907.

672 JAMESON, William Scarlett, 1899- . *The Most Formidable Thing: The Story of the Submarine from its Earliest Days to the End of WW I*. London, Hart-Davis, 1965. 280 pp., illus., plates, maps.

673 ________. *Submariners, V.C.* Foreword by G. W. G. Simpson. London, Davies, 1962. 208 pp., ports, maps, diagrams, bibliog.

The stories of the 14 Royal Navy submariners who won Victoria Crosses in World Wars I and II.

674 JANE, Frederick Thomas, 1865-1916. *The Imperial Japanese Navy*. . . . London, Thacker, 1904. Calcutta & Simla, Thacker, Spink & co., 1904. 410 pp., frontis., illus., plates, ports, maps, plans. LC 5-4260 rev.

Covers period of Russo-Japanese War of 1904-05. More than 80 of the illustrations are from sketches and drawings done by Japanese artists, or from official photographs.

675 ________. *Imperial Russian Navy: Its Past, Present and Future.* London, NY, etc., Thacker, 1899. 755 pp., frontis., illus., plates. LC 13-6814.

The Russian Fleet as it was on the eve of the Russo-Japanese War.

676 ________. *The Imperial Russian Navy: Its Past, Present and Future.* New and revised ed. London, Thacker. 1904. 736 pp., frontis., illus., plates, ports, maps, diagrams, facsimiles. LC 4-14871.

677 ________. *The Torpedo in Peace and War.* London, Thacker, 1898. 163 pp., illus.

Japanese Maritime Defense Force. See entry 35.

Japanese Navy. See entries 674, 994, 1252, 1447.

The Japanese Navy—At the End of WW II. See entry 451.

The Japanese Navy in WW II. See entry 933.

Japanese Submarines. See entries 35, 450, 451, 512, 578, 580, 621, 1486, 1487.

678 JEFFERIS, Roger and Kendall McDONALD. *The Wreck Hunters.* NY, Barnes, 1967. London, Harrap, 1967. 255 pp., illus. LC 67-13132. (American Ed.) LC 66-72239 (English Ed.)

679 JELLICOE, John Rushworth Jellicoe, 1st Earl., 1859-1935. *The Crisis of the Naval War.* London, NY, etc. Cassell, 1920. 331 pp., frontis., plates, maps. LC 20-22471 Revised.

680 ________. *The Crisis of the Naval War.* NY, Doran, 1921. 331 pp., frontis., plates, maps. LC 21-1388.

681 ________. *Erinnerungen; Zwischen Skagerrak und Scapa Flow.* Translated by Johannes Spiess. Berlin, Vorhut, 1937. 247 pp., illus., ports, maps. LC 50-48996.

German translation of *Crisis of the Naval War.*

682 ________. *Il Pericolo Sottomarino: La Politica dell' Ammiragliato nel 1917.* Translated by Fausto Leva. Rome, Instituto Poligrafico dello Stato, Libreria, 1936. 132 pp., tables. LC 36-16277.

683 ________. *The Submarine Peril: The Admiralty Policy in 1917.* London, Cassell, 1934. xvi, 140 pp., frontis, plates, ports. LC 35-4786.

684 ________. *Der U-Boot-Krieg, Englands Schwerste Stunde.* Translated by Johannes Spiess. Berlin, Vorhut/Schlegel, 193? 229 pp., plates, maps. LC 50-33588.

German translation of *The Submarine Peril.*

Jellicoe, John Rushworth, 1st Earl, 1859-1935. See also entry 484.

685 JENKINS, E. H. *A History of the French Navy.* London, Macdonald and Jane's, 1973. 364 pp., illus., maps, bibliog.

The Jesse James of the Java Sea. See entry 189.

686 JEWELL, Norman Limbury Auchinleck. *Secret Mission Submarine.* As told to Cecil Carnes. NY, Ziff-Davis, 1945. Toronto, Ambassador, 1945. 159 pp. LC 45-2988.

Re mission of H.M.S. Seraph.

Jewell, N. L. A. See also entry 1047.

687 JEWETT, H. K. *Submarine Adventures of Little Jinks.* Illustrated by Herbert Rasche. Los Angeles, Wetzel, 1930. 79 pp., illus. LC 30-31594.

Humor.

Jim Thorne's Guide to Adventure. See entry 1245.

Jimmy-the-One. See entries 1473, 1474.

688 JOHNSON, E. E. *Pressure Tank and Instrumentation Facilities for Studying the Strength of Vessels Subjected to External Hydrostatic Loading.* Research and development report. Presented at the annual meeting of the Society for Experimental Stress Analysis in Philadelphia. September 1954. Washington, 1956. David W. Taylor Model Basin Report 979, pp. 124-139. LC 56-62384.

Johnson, John E. See entry 1102.

Johnson Sea-Link. See entry 783.

689 JONES, Thomas Michael. *Watchdogs of the Deep: Life in a Submarine During the Great War.* Foreword by W. T. Randle Ford. Sydney, Angus, 1935. London, Australian Book Co., 1935. 224 pp.

Jouan, René, 1894- . See entries 375, 1167.

690 *Journal des Scavans, (Annees) 1665-1910.* Annals of diving and underwater work. Amsterdam, Menno Hertzberger, 1968.

A facsimile reprint, with 144 plates, was issued by Hertzberger bound in 210 volumes.

691 *Journal of Ocean Technology.* 1966-

Name was changed to *Marine Technology Society Journal.*

The Journal of Submarine Commander von Forstner. See entry 417.

692 JULIER, Ferenc, 1878- . *A Búvárnaszád Háború.* Budapest, Victorisz, 1940. 52 pp. LC 50-53064.

World War I submarine operations.

693 JUNG, H.A.K. *Krieg Unter Wasser.* Oldenburg, Stalling, 1939. 222 pp., illus.

Jusserand, Jean A.A.J., 1855- . See entry 1435.

K

K-XIII (Dutch submarine). See entry 781.

The K Boats: A Dramatic First Report on the World's Most Calamitous Submarines. See entry 384.

Kaiten. See entries 1486, 1487.

The Kaiten Weapon. See entry 1486.

Kampf und Untergang der Kriegsmarine. See entry 90.

Das Kampfschicksal der Deutschen Flotte im Weltkrieg. See entry 418.

694 KAPITAN, Gerhard and John HUSTON. *A Bibliography of Underwater Archaeology.* Chicago, Argonaut, 1967. 144 pp. LC 66-19178.

Bibliographical survey of literature and reports of underwater archaeology in the Mediterranean, Black Sea, and Northern Europe.

Kapitänleutnant Otto Steinbrinck. See entry 602.

Karig, Walter, 1898- . See entry 253.

695 KARNEKE, Joseph Sidney. *Navy Diver* as told to Victor Boesen. NY, Putnam, 1962. Toronto, Longmans, 1962. 256 pp., plates. LC 62-7348.

Same, with Victor Boesen. Chicago, Hale, 1962. 190 pp.

696 KASER, Arthur LeRoy, 1890- . *The Stolen Submarine: A Burlesque in One Act.* Minneapolis, Northwestern Press, ca. 1944.

Kashiwagi, Isami. See entry 712.

Kasten, Fritz, d. 1931. See entry 441.

697 KEEBLE, Peter. *Ordeal by Water.* Garden City, Doubleday, 1958. 216 pp., illus. LC 58-7360.

British hard-hat salvage diving on the U-307. WW II in the Mediterranean.

Keller, Hannes. See entry 1419.

698 KELLEY, D.M. *22 Cells in Nuremberg.* NY, 1947.

"With a chapter on Doenitz by the American psychiatrist to the Nuremberg jail."—Ladislas Farago.

699 KEMP, Peter Kemp. *H.M. Submarines.* London, Jenkins, 1952. Toronto, Longmans, 1952. 224 pp., illus. LC 53-21602.

Kendall, Henry Way, 1926- . See entry 1101.

Kenyon, Norman. See entry 1134.

Kerr, Mark. See entry 866.

Keyes, Sir Roger. See entries 358, 359.

Kiel, Kurt. See entry 1110.

The Killing Time. The U-Boat War 1914-18. See entry 508.

Kimball, William Wirt, 1848-1930. See entry 175.

King, Bert T. See entries 1348, 1351.

700 KING, William, 1910- . *The Stick and the Stars.* NY, Norton, 1958. London, Hutchinson, 1958. 192 pp. LC 58-7178. (58-2560)

Kinkead, Beatrice. See entry 1494.

Kinsey, Jack L. See entry 1344.

701 KIPLING, Rudyard, 1865-1936. *The Fringes of the Fleet.* Garden City, Doubleday, Page, 1915. 122 pp.
Includes anecdotes of WW I British submarine service.

702 ________. *Il Mestiere, Storie di Sottomarini Britannici.* . . . translated by Ettore Bravetta. London, Darling, 1916. 54 pp. LC 38-16702.

703 ________. *Tales of "The Trade".* Garden City, Doubleday, Page, 1916- __v. LC 16-14356.

704 ________ *Die Taten des Tauchboots: Erzahlungen aus dem "Handelstrieb".* London, Darling, 1916. 55 pp. LC 25-21221.

705 KIRCHHOFF, Hermann, 1851 , ed. *Otto Weddigen und Seine Waffe. Aus Seinen Tagebuchern und Machgelassenen Papieren, Unter Mitwirkung der Familie.* Berlin, Marinedank, 1915. 157 pp., frontis., illus., port. (Band 2 of Unsere Seehelden series).

706 ________ and A. MARCUSE. *U-Boot und Luftboot.* Berlin, Collignon, 1915. (Deutsche Kraft. Heft 18). 30 pp., illus., diag.

Kirchhoff, Hermann, 1851 see also entry 1441.

707 KIRCHNER, Johann. *Das U-Boot Bei der Arbeit: Seine Technik und Wirkungsweise in Wort und Bild.* Berlin,

Mittler, 1917. 86 pp., diagrams, plates.

708 KITSAK, N. A. *Rezka Metallov Pod Vodoi i Raschistka Rusel.* 1950. 50 pp., illus., bibliog. LC 55-35571.

Technical manual on underwater welding and cutting.

Kitty Hawk to Sputnik to Polaris. See entry 1471.

709 KLADO, Nicholas Lavrentievich, 1861- . *The Russian Navy in the Russo-Japanese War.* Translated by Rene Marchand. London, Hurst & Blackett, 1905. 281 pp., frontis., plates. LC War 8-32.

Klaxon. *See* Bower, John Graham. and entry 135.

710 KLEINLEIN, Gerhard. *Bezwungene Tiefen.* Berlin, Sportverlag, 1956. 186 pp., illus. LC 57-16424.

Skin and scuba diving for sport.

K-Men: The Story of the German Frogmen and Midget Submarines. See entry 88, 89.

Knight, Austin Melvin, 1854- . See entry 1377.

711 KNOX, Philander Chase, 1853- . *Address of . . . at a Patriotic Mass Meeting in Exposition Music Hall, Pittsburgh, Pa., March 31st, 1917.* Phila., Allen, Lane and Scott, 1917. 16 pp. LC 17-21848.

712 KNOWLTON, William. *Let's Explore Beneath the Sea.* Foreword by Jack Ackerman. Illustrated by Isami Kashiwagi. NY, Knopf, 1957. 139 pp., illus., bibliog. LC 57-5259.

713 KOEBSELL, Eberhard, 1906- . *U-31, das Schiff aus dem Jenseits.* Berlin, Verlag Die Wehrmacht, 1937. 187 pp., plates. LC A F 47-6291.

Koebsell, Eberhard, 1906- . See entry 737.

714 KOENIG, Paul. *Die Fahrt der Deutschland.* Berlin, Ullstein, 1916. 158 pp., illus.

The German merchant submarine *Deutschland* voyaged from Bremen to Baltimore with a load of dye stuffs, and returned to Germany with a load of drugs. This was previous to the entry of the USA into World War One.

715 ————. *Die Fahrt der Deutschland: das Erste Untersee-Frachtschiff.* NY, Hearst's International Library, 1916. xii, 254 pp., frontis., port. and 27 photos.

716 ________. *El Primer Viaje del Submarino Mercante Deutschland.* Buenos Aires, Schneider, 1918. 142 pp., frontis., plates, ports.

717 ________. *Voyage of the Deutschland: The First Merchant Submarine.* NY, Hearst's International Library, 1916. 247 pp., frontis., illus., ports.

Koenig, Paul. See also entry 1439.

Kogrund Passage. See entry 520.

718 KOHL, Jessie W. *Summaries of Research Reported on During 1966.* Groton, Conn., U.S. Naval Medical Center, Submarine Base, Feb. 15, 1967. Special Report No. 67-4. (D 206.10/3:67-4) viii, 31 + 2 pp.

Kolands, G. N. See entry 503.

719 KOLBASEV, Sergei A. *Voenno-Morskoi Flot.* Leningrad, 1926.

720 KOLYSHKIN, Ivan Aleksandrovich, 1902- . *Submarines in Arctic Waters.* Translated by David Skyirsky. Moscow, Progress Pubs., 1966. 253 pp., illus.

721 KONDRATEV, N.D. *Sovestsky Voenno-Morskoi Flot.* (The Soviet Navy). Moscow, 1957.

Konstruktsiia Podvodnykh Lodok. See entry 995.

Korabli Morskikh Glubin. See entry 1128.

722 KORGANOFF, A. *Les Sous-Marins.* Paris, Hachette, 1963.

723 KORNIENKO, Daniil Iosifovich. *Flot Nashei Rodiny.* (Our Homeland's Fleet). Moscow, 1957.

724 ________. *SSSR—Velikaia Morskaia Derzhava.* (The USSR—A Great Sea Power). Moscow, 1950.

725 ________. and N. MILGRAM. *Voenno-Morskoi Flot Sovetskoi Sotsialisticheskoi Dershavy.* (The Navy of the Soviet Socialist State). Moscow, 1949.

Kovalik, Nada. See entry 726.

726 KOVALIK, Vladimir, 1929- and Nada. *Undersea World of Tomorrow.* Illustrated by George Solonevich. NY, Prentice-Hall, 1969. 51 pp. LC 69-12365.
Juvenile

727 KRAMSTA, Joachim, 1892-1917. *Aus dem Logbuch des I. Wachoffiziers, U66: Auszuge aus Briefen und Tagebuchblattern.* Hanover, Verlag Industrie-und Handelsdienst, 1931. 189 pp., plates, port., facism. LC 51-51607.

Krasny Morskoi Flot. See entry 495.

Krawczyk, John. See entries 22, 23.

Kretschmer, Otto. See entries 1041, 1042, 1045, 1046, 1048.

Kreuzerfahrten und U-Boots Taten. See entry 502.

Der Krieg in der Nordsee. See entry 535.

Krieg Unter Wasser. See entry 693.

728 KRIEGSARCHIV. VIENNA. *Report on Details of the Austro-Hungarian Submarines: Specially prepared for the Information of Authors.* Vienna, 1919.

729 *Kriegstagebuch des Befehlshavers der U-Boote.*
"War diaries of the BdU, Commander, U-Boats (Doentiz)"—Ladislas Farago.

730 KRUSKA, Emil. *Walter U-Boote.* Munich, Lehmann, 1969. 216 pp., illus., drawings, deck plans, diagrams.
A technical treatise on all of the designed and projected Walter nydrogen peroxide, high speed U-boats.

731 KUENNE, Robert Eugene, 1924- . *The Attack Submarine: A Study in Strategy.* New Haven, Yale University Press, 1965. 215 pp., maps. LC 65-22329.

732 ———. *The Polaris Missile Strike: A General Economic Systems Analysis.* Columbus, Ohio State University Press, 1967. xxv, 434 pp. LC 66-10715.

733 KUESTER, Julius, ed. *Des U-Boot als Kriegs-und Handelsschiff: Die Technische Entwicklung und Anwendung der Tauchboote, deren Motoren, Bewaffnung und Abwehr. . . .* Berlin, Klasing, 1918. 166 pp., illus., diagrs., tables.

Die Kuhne Fahrt der Deutschland. See entry 143.

734 KUNSTMANN, Emil, ed. *Handels-U-Boot Deutschland.* Leipzig, Vogel & Vogel, 1916? 64 pp., frontis. port, illus., plates.

735 KURZAK, K. H. and Eberhard ROESSLER. *Unterseeboote und Torpedos mit Kreislaufantrieb.* Kiel, Privately Printed,

1969. 162 pp., illus., diagrams.

A technical treatise on closed cycle submarine propulsion system.

736 KUZNETSOV, I. I. *Rukovodstvo Diia Vodolaza.* Moscow, 1954. 182 pp., illus. LC 55-27680.

L

L-9, USS (SS 49). See entries 1239, 1240.

737 LAAR, Clemens, *pseud. Das Geister U-Boot: U-31 das Schiff aus dem Jenseits.* Berlin, Wehrmacht Verlag, 1939. 187 pp., illus.

Author's real name is Eberhard Koebsell, 1906-

Laar, Clemens, *pseud.* See also entry 713.

738 La BRUYÈRE, René, 1875- . *Notre Marine Marchande Pendant la Guerre.* Paris, Payot, 1920. 384 pp., tables. LC 22-8125.

The French Merchant Marine during WW I., and the submarine war.

Laconia, SS. See entry 956.

The Laconia Affair. See entry 956.

739 LAKE, Simon, 1866-1945. "Safe Submarine Vessels and the Future of the Art." London, Institution of Naval Architects, 1907. 28 pp., diagram, plates.

740 ________. *Submarine: The Autobiography of Simon Lake as Told to Herbert Corey.* NY, London, Appleton-Century, 1938. xiv, 303 pp., frontis., illus., ports. LC 38-27651.

741 ________. *The Submarine in War and Peace: Its Developments and its Possibilities.* Phila., & London, Lippincott, 1918. xi, 301 pp., frontis., port., illus., plates, chart. LC 18-18750.

Lake, Simon, 1866-1945. See also entries 362, 363, 364, 462, 742-745, 907, 1419.

742 LAKE SUBMARINE COMPANY. *The Submarine Boat Argonaut. What She Was Built For and What She Has Accomplished.* NY, 1900. 16 pp., illus.

Lake Submarines. See entries 746, 1137.

743 LAKE TORPEDO BOAT COMPANY. *The Development of the Lake Type Submarine Torpedo Boats.* Bridgeport, Conn., 1913. 27 pp., illus.

744 THE LAKE TORPEDO BOAT COMPANY. *Submarine Torpedo Boats.* NY, The Lake Torpedo Boat Co., 1901. 31 pp., illus. LC 2-28711.

745 ________. *Under-Water Torpedo-Boats: The Submarine versus the Submersible, Their Merits and their Menace.* Bridgeport, The Lake Torpedo Boat Co., 1906. 116 pp., frontis, illus. LC 17-20872.

Lake Torpedo Boat Co. See also entries 1143, 1288.

Lambertsen, C. J. See entries 1211, 1212.

746 LAN, Luis Angel, 1867- . *Submarines Lake and Holland. Report of Commander L. A. Lan, Argentine Navy.* Translated by M. E. Beall. Buenos Aires, December 10, 1904. U.S. Navy Dept. *Review of Naval Publications.* No. 87. 40 pp., plates. LC 5-15861.

Lancaster, William Joseph Cosens. See entry 239.

747 LANDQUIST, Daniel, 1891- . *Den Tyska Ostasiatiska Eskaderns Öden Under Världskriget 1914-15.* Stockholm, Marinlitteratur-föreningen, 1915. 104 pp., illus., map. LC 21-15690.

748 LANGSDORFF, Werner von, 1899- . *U-Boote am Feind: Fünfudviersig Deutsche U-Boot-Fahrer Erzählen.* Gütersloh, Bertelsmann, 1943. 367 pp., plates, ports. LC A F 47-6852.
World War One German submarine operations.

Lanier, Henry Wysham, 1873- . See entries 100, 101.

Lanphear, R. E. See entry 1393.

Lanzamiento Nocturno Presuroso. See entry 476.

Lardner, Rex. See entry 1407.

Larsen, Egon, *pseud.* See entry 768.

749 LARSON, Howard E. *A History of Self-Contained Diving and Underwater Swimming.* Washington, Committee on Undersea Warfare. National Academy of Sciences. National Research Council, 1959. (Pub. No. 469). 50 pp., illus. LC 59-60059.

750 LASSEN, Ernst. *Handels-U-Boot Deutschland Fahrt Nach Amerike.* Siegen, Montanus, 1916. 111 pp., frontis., plates, port.

The Last Cruise. See entry 762.

Last Days of the German Fleet. See entry 439.

The Last Fathom. See entry 177.

The Last Voyage of the Lusitania. See entry 615.

751 LATIL, Pierre de, 1905- and Jean RIVOIRE, 1929- . *Man and the Underwater World.* Translated by Edward Fitzgerald. London, Jarrolds, 1956. 400 pp., illus. LC 56-3095.

752 ________. Same title. NY, Putnam, 1956. 400 pp., illus., LC 56-7203.

753 LAUBEUF, Alfred Maxime, 1864- . *Les Sous-Marins Allemands: Les Divers Types, Leurs Caracteristiques, Leur Action Pendant la Guerre, Moyens Employes Pour les Combattre.* Paris, Delagrave, 1920. 40 pp., tables, diagrams, plate.

754 ________. *Sous-Marins et Submersibles, Leur Dévelopement, Leur Rôle dans la Guerre, Leur Rôle dans L'Avenir, Les Sous-Marins Allemands.* Paris, Delagrave, 1915. 100 pp., illus., plates, bibliography. LC 15-18977.

An important source book.

755 ________. *Sous-Marins, Torpilles et Mines,* with Henri Stroh. Paris, Bailliere, 1923. 810 pp., illus., plates, tables, diags., bibliog. LC PO 25-40 Revised.

The Laughing Cow. See entry 856.

756 LAURENS, Adolphe, 1875- . *Le Blocus et la Guerre*

Sous-Marine (1914-1918) Paris, Colin, 1924. 215 pp., map, diagram, bibliog. LC 24-10319.

757 ________. *Introduction à L'Étude de la Guerre Sous-Marine.* Paris, Challamel, 1921. 233 pp., plates, diagrams, tables. LC 21-13244.

Laurens, Adolphe, 1875- . See also entry 1006.

Laurenti, G. See entry 1001.

758 LAURIAT, Charles Emelius, Jr. *Lusitania's Last Voyage.* Boston, Houghton Mifflin, 1915. 158 pp., frontis., plates. LC 15-25328

759 La VALLEE, David. *Event 1000: A Novel.* NY. Holt, Rinehart and Winston, 1971. 279 pp. ISBN 0-03-085969-7. LC 73-138881.

Novel about a sunken submarine. "Event 1000" is USN code for sub sunk.

760 LAWRENZ, Hans Joachim. *Die Enstehungsgeschichte der U-Boote.* Munich, Lehmann, 1968. 147 pp., illus., diagrams.

History of submarines from earliest times to the present.

761 LEBEDSKOI, G. M. *Torpeda i Mina.* Moscow, 1949. 102 pp., illus. LC 50-29916.

Lecture on Drifting and Automatic Movable Torpedoes. . . . See entry 49.

Lecture on Submarine Boats and Their Application to Torpedo Operations. See entry 50.

762 LEDERER, William J., 1912- . *The Last Cruise: Story of the Sinking of the Submarine USS Cochino.* NY, Sloane, 1950. 110 pp.

Has sailing lists of both the Tusk (SS 426) and the Cochino (SS 345)

Lee, Ezra, 1749-1821. See entries, 2, 3, 617.

Lee, Manning de V. See entry 379.

767 LEE, Owen Scott, 1926- . *The Complete Illustrated Guide to Snorkel and Deep Diving.* Foreword by Jacques Yves Cousteau. NY, Doubleday, 1962. Toronto, 1962. 467 pp. LC 62-19793.

"The art and techniques, pleasures and profits, equipment, spearfishing skills, and USN safety standards."

768 LEHRBURGER, Egon, 1904- . *Men Under the Sea,* by Egon Larsen, *pseud.* London, Phoenix House, 1955. 223 pp., illus. LC 56-1068.

Skin and scuba diving.

763 LEIGHTON, John Langdon. *Simsadus: London; The American Navy in Europe.* NY, Holt, 1920. 169 pp., frontis., plates, port, diagrams. LC 20-9639.

WW I, USN submarines in European theatre of operations.

Leiper, Henry Smith, 1891- . See entry 916.

Le Loup de L'Atlantique. See entry 1044.

764 Le MASSON, Henri. *Du Nautilus (1800) au Redoutable.* Paris, Presses de la Cite, 1969. 455 pp., illus.

History of French submarines from the Nautilus of 1800 to the Redoutable of 1969.

765 ________. *Histoire du Torpilleur en France 1872-1940.* Paris, Academie de Marine, 1967. 370 pp., illus.

766 ________. *La Marine de Guerre Moderne. (Porte-Avions, Sous-Marins, Escorteurs)* Paris, Horizons de France, 1950. 164 pp., illus.

767 Listed out of sequence see above under LEE.

768 Listed out of sequence see above under LEHRBURGER

Le May, Curtis E. See entry 794.

769 LENT, Henry Bolles, 1901- . *Submariner: The Story of Basic Training at the Navy's Famed Submarine School.* NY, Macmillan, 1962. 182 pp., illus. LC 62-8557.

Illustrated with General Dynamics photos and official USN photos.

770 LENTON, Henry Trevor. *American Submarines.* (Navies of the Second World War). NY, Doubleday, 1973. 128 pp., illus. ISBN 0-385-04761-4. LC 72-86661.

771 ________. *American Submarines.* (Navies of the Second World War). London, MacDonald, 1973. 128 pp., illus. ISBN 0-356-04190-5. LC 73-177320.

772 ________. *British Submarines.* (Navies of the Second World War). NY, Doubleday, 1972. 160 pp., illus. LC 79-187126.

773 ________. *German Submarines.* (Navies of the Second World War). NY, Doubleday, 1966. London, MacDonald, 1966. 2 volumes. v.I: 126 pp. v.II: 128 pp. Illus.

Leonard, John M. See entry 1387.

774 LePOTIER, A. *Sous La Banquise.* Paris, France-Empire, 1951.

Le PRIEUR, Yves. See entry 1018.

Lerner Marine Laboratory. See entry 1218.

L'Estrange, C. J. See entry 1191.

Let's Explore Beneath the Sea. See entry 712.

Let's Go Aboard an Atomic Submarine. See entry 560.

Letter to the Secretary of the Navy . . . on the Practical Use of the Torpedo. See entry 453.

Letters . . . on Sub-Marine Navigation and Attack. . . . See entry 454.

Leuchtman, Eugene. See entry 482.

Leva, Fausto. See entry 682.

775 LEVY, Hermann, 1881- . *1917, Februar. 1918: Der Sinn des U-Bootkrieges, nach Einmen Jahr der Seesperre.* Charlottenberg, Raben, 1918. 59 pp. LC 24-5038.

776 LEWELLEN, John. *The Atomic Submarine.* NY, Crowell, 1954. 134 pp., illus. LC 54-5540.

777 LEWIS, David D. *The Fight for the Sea: The Past, Present and Future of Submarine Warfare in the Atlantic.* Foreword by ADM. Jerauld Wright. Cleveland, World, 1961. Toronto, Nelson, Foster, & Scott, 1961. 350 pp., bibliog. LC 61-5991.
An important study. Good bibliography.

778 LEYLAND, Eric, 1911- . *Crash Dive.* Illustrated by B. Gerry. London, E. Ward, 1961. NY, British Book Service, 1961. 128 pp., plates.

Leyland, John. See entry 463.

779 L'HERMINIER, Jean, 1902- . *Casabianca: The Secret Missions of a Famous Submarine.* Translated by Edward Fitzgerald. London, Muller, 1953. Toronto, Saunders, 1953.

243 pp., plates.

La Liberté des Mers. See entry 1413.

Libro Strumenti. . . . See entry 797.

Life in a Submarine. See entry 9.

The Life of Robert Fulton. See entry 237.

Lilley, George Leavens, 1859- . See entry 1297.

780 LIMBURG, Peter R., 1929- and James B. SWEENEY. *Vessels for Underwater Exploration: A Pictorial History.* NY, Crown, 1973. 154 pp., illus. with 100 photos. ISBN 0-517-50534-7. LC 73-78878.

Slanted toward a juvenile audience; valuable for the pictures.

781 LINDEN, C. van der and M. S. WYTEMA. *Met Hr. Ms. K.XIII naar Nederlandsch-Indie; een Onderzeebootocht van Meer den 20,000 Zeemijlen.* Amsterdam, Scheltens & Giltay, 1927. 252 pp., frontis., plates, maps.

Voyage of 20,000 miles made by the Dutch submarine K-XIII from Holland to the Netherlands Indies.

Lindley, Denver. *See entry 972, 158a.*

782 LINK, Marion Clayton. *Sea Diver. A Quest for History Under the Sea.* NY, Holt, Rinehart & Winston, 1959. 333 pp., illus., bibliog. LC 59-5333.

Underwater archaeological expedition in Caribbean. Diving. Columbus ship.

783 ________. *Windows in the Sea.* Foreword by I. Eugene Wallen. Washington, Smithsonian Institution Press, 1972. 192 pp., illus., partly colored. ISBN 0-87474-130-0. LC 72-93801.

"The step-by-step evolution of an underwater research vessel, the *Johnson Sea-Link,* is told in a readable, knowledgeable way by an eye witness. . . . To answer the needs of research biologists, Edwin A. Link developed a submersible with . . . a transparent plastic bubble permitting panoramic vision and a diver lock-out which allows the observer to emerge on the sea floor . . ." publisher's listing.

784 LIPSCOMB, F. W. *The British Submarine.* NY, Macmillan, 1954. London, A&C Black, 1954. 269 pp., illus., sectional drawings, diagrams, maps.

Covers history of British submarines from their beginnings through World War Two, with speculation on the future.

785 ________. *Historic Submarines.* Illus. by Malcom McGregor. NY, Praeger, 1970. London, Hugh Evelyn, 1970. 36 pp. text, plus 16 color plates.

Finely executed color plates begins with the 1898/99 French *Narval,* includes submarines of various nations, and ends with USN atomic Polaris missile submarine *George Washington.*

786 *A List of Neutral Ships Sunk by the Germans from August 8, 1914 to April 26, 1917.* London?, 1917. 32 pp. LC 17-22322.

The Little David, C. S. S. David. See entry 1149.

The Living Sea. See entry 258.

Lleget Colomer, Mario. See entry 1029.

787 LLUZAR RODRIGO, José. *Los Buques Submarinos Actuales, Su Construcción, Armamento, Maniobras, Táctica y Empleo en la Defensa de las Costas y en la Guerra Naval.* Madrid, Bailly-Bailliere, 1916. 440 pp., illus., plates, diagrams, bibliog. LC 43-41011.

788 LOCKWOOD, Charles Andrews. 1890-1967. *Down to the Sea in Subs.* NY, Norton, 1967. 376 pp., illus. LC 66-12798.

Memoirs of the man who was Commander of Submarines, Southwest Pacific during WW II. Covers forty years of American submarine history.

789 ________ and Hans Christian ADAMSON, 1890 . *Hell at 50 Fathoms.* Phila., Chilton, 1962. 299 pp., illus. LC 62-8342.

Submarine disasters, salvage and rescue.

790 ________ & ________. *Hellcats of the Sea.* Foreword by ADM. Chester W. Nimitz. NY, Greenberg, 1955. Toronto, Ambassador, 1955. Phila., Chilton, 1955. 335 pp., plates. LC 55-10961.

Operation Barney and USN submarines in Pacific during WW II.

791 ________. *Sie Jagten Nippons Flotte—die Amerikanischen U-Boote im Pazifik 1941-1945.* Krefeld, Ruhl, 1963.

German translation of *Sink 'Em All* edited by Gunter Schomaekers. American foreword by ADM. Chester W. Nimitz. German foreword by ADM. Karl Doenitz.

792 ________. *Sink 'Em All: Submarine Warfare in the Pacific.* Foreword by ADM Chester W. Nimitz. NY, Dutton,

1951. Toronto, Smithers & Bonellie, 1951. 416 pp., illus. LC 51-9835.

USN submarine operations against the Japanese in the Pacific during WW II.

793 ________ & ________. *Through Hell and Deep Water: The Story of the Navy's Deadly Submarine, the USS Harder, under the Command of Sam Dealey, Destroyer Killer.* NY, Greenberg, 1956. Phila., Chilton, 1956. xiv, 317 pp., illus. with photos. LC 56-11388.

Appendix lists all persons who ever served aboard the USS *Harder.* (SS 257).

794 ________ & ________. *Zoomies, Subs and Zeroes: Heroic Rescues in World War Two by the Submarine Lifeguard League.* Forewords by C. A. Pownall and Curtis E. LeMay. NY, Greenberg, 1956. Phila., Chilton, 1956. xv, 301 pp., illus., map endpapers. LC 56-7976.

Lockwood, Charles Andrews, 1890-1967. See also entries 601, 1185.

Lody, Karl Hans, d.1914. See entry 713.

The Log of a U-Boat Commander. See entry 574.

Log of the Flying Fish. See entry 239.

795 LOHMANN, W. and H. H. HILDEBRAND. *Die Deutsche Kriegsmarine 1939-1945.* Bad Nauheim, Hans-Henning, 1956.

London. See entry 763.

796 LONDON, SCIENCE MUSEUM. *Naval Mining and Degaussing: An Exhibition of Representative Britsh and German Naval Mining and Degaussing Material Used During the War, 1939-1945.* London, HMSO, 1946. vii, 27 pp. LC 49-44952.

Long, Frank D. See entry 866.

Long, R. O. B. See entries 431, 432.

A Long Line of Ships. See entry 798.

Lonsdale, Rupert. See entry 1429.

Look at Submarines. See entry 1488.

Loomis, F. Kent. See entry 314.

Lord of the Seas. See entry 1191.

797 LORINI, Buonaiuto. *Libro Strumenti ne Quaki Possono Star Gli Huomini Sotto Acqua.* Venice, 1609.

Book on fortifications and anti-submarine defenses published originally in 1607.

Lost Men of American History. See entry 617.

The Lost Sentinel. See entry 563.

The Lost Ships. See entry 1249.

798 LOTT, Arnold Samuel, 1912- . *A Long Line of Ships: Mare Island's Century of Naval Activity in California.* Annapolis, U.S. Naval Institute, 1954. 268 pp., illus. LC 55-328.

A centennial history of Mare Island Navy Yard.

Les Loups de L'Amiral. See entry 923.

799 LOW, Archibald Montgomery, 1888- . *Mine and Countermine.* Foreword by T. B. Thompson. NY, Sheridan House, 1940. 224 pp., plates. LC 40-27812.

800 ________. *The Submarine at War.* Foreword by ADM William V. Pratt. NY, Sheridan House, 1942. 305 pp., plates. LC 42-16172.

801 LOWTHER, George Francis, 1913- . *Submarine: A One Act Drama.* NY, French, 1946. 17 pp. LC 46-21657.

802 LUEDDE-NEURATH, W. *Regierung Doenitz. Die Letzen Tage des Dritten Reiches.* Goettingen, 1951.

Doenitz's last days in power by a member of his staff.

803 LUMIERE, Cornel. *Beneath the Seven Seas.* London, Hutchinson, 1956. 240 pp., illus. LC 56-58410.

Diving and salvage. Description of underwater world & marine fauna.

804 LÜTZOW, Friedrich, 1881- . *De Strijd Onder Water, Mijnen en Onderzeebooten als Blokkadebrekers.* Foreword by M. A. Cageling. Amsterdam, Scheltens & Giltay, 1940. 56 pp., illus. LC 54-51871.

Dutch translation of *Mit U-Boot und Minen Gegen Englische Hunger-Blockage.* WW I.

805 ________. *Unterseebootskrieg und Hungerblockage.* Berlin-Dahlem, Meyer, 1921. 51 pp., bibliographical footnotes. LC A 21-1144 rev.

806 LUSAR, Rudolf. *Die Deutschen Waffen und Geheimwaffen.* 5th ed. Munich, Lehmann, 1964. 379 pp., illus.

Describes German weapons and secret weapons of WW II. Includes submarines.

Lusitania (steamship). See entries 77, 341, 366, 448, 615, 645, 758, 1135, 1321.

The Lusitania. See entry 1135.

The Lusitania Case. See entry 341.

Der Lusitania-Fall und de Deutsche Wissenschaft. . . . See entry 77.

Lusitania's Last Voyage. See entry 758.

Lyon, F. H. See entry 827.

M

807 MAAS, Peter. *The Rescuer.* NY, Harper, 1967. 239 pp., illus., bibliog. LC 67-23048.

Squalus disaster and the role of Charles B. "Swede" Momsem in the rescue and salvage operations.

808 McCANN, Hugh Wray, 1928- . *Utmost Fish!* NY, Simon & Schuster, 1965. 384 pp. LC 65-12951.

Fiction

809 McCOY, Jack. *The Big Book of Real Submarines.* NY, Grosset & Dunlap, 1955. (Big Treasure Books juvenile series) unpaged, illus. LC 55-4105.

810 ________. *The Book of Submarines.* NY, Grosset & Dunlap, 1966. unpaged, illus. LC 67-454.

Juvenile.

McCue, Elizabeth C. See entry 1323.

811 MccGWIRE, Michael, ed. *Soviet Naval Developments:*

Context and Capability. NY, Praeger, 1973. 560 pp., illus.

Papers of a seminar held at Dalhousie University in Halifax, N.S. during October, 1972.

MacDonald, James. See entry 1493.

812 McDONALD, Kendall and Phil SMITH. *Spearfishing in Britain*. (Sportshelf Book) New Rochelle, Soccer, 1964. London, Paul, 1964. 159 pp., illus., plates. LC 64-6344.

McDonald, Kendall. See also entry 678.

813 MacDONNELL, James Edmond, 1917- . *Subsmash*! London, Constable, 1960. 288 pp.

Subsmash is code for sunken submarine. James MacNeil is *pseud.* of MacDonnell.

McGregor, Malcolm. See entry 785.

814 MacHARDY, Charles. *Send Down a Dove*. NY, Coward-McCann, 1968. First published in England. 351 pp. LC 68-23370.

Documentary fiction re British submarine service during WW II. Fictional submarine HMS *Scorpion*.

De la Machine Infernale Maritime. See entries 452, 459.

815 MacINTYRE, Donald. [George Frederick Wyville], 1904- . *Fighting Under the Sea*. NY, Norton, 1966. 174 pp., illus. with photos. LC 66-18074.

Popular anecdotal history of submarine warfare from 1578 through WW II.

816 ________. *Fighting Under the Sea*. London, Evans, 1965. 174 pp., plates. LC 66-57738.

817 ________. *U-Boat Killer*. Foreword by R. B. Carney. NY, Norton, 1956. 239 pp., plates. LC 57-5989.

Same. London, Weidenfeld, 1956. NY. Ambassador, 1956. 179 pp., plates. LC 56-14416.

818 MacINTYRE, Donald and Basil W. BATHE. *Man-of-War: A History of the Combat Vessel*. NY, McGraw, 1970. 273 pp., illus., bibliog. LC 68-23576.

International account of development of warships spanning 6000 years. Covers the salient points of the history.

819 McKEE, Alexander, 1918- . *Black Saturday*. NY, Holt, 1960. 211 pp., plates. LC 60-7335.

Same with subtitle: *The Tragedy of the Royal Oak*. London, Souvenir, 1959. 220 pp., illus., plates, maps. LC 60-17840.

Gunther prien's penetration into Scapa Flow in his U-47 and the sinking of HMS *Royal Oak*.

820 ________. *History Under the Sea*. London, Hutchinson, 1968. 342 pp., illus., maps. LC 74-353114.

American edition pub. in NY, Dutton, 1969. LC 69-20307. Underwater archaeology.

821 MacLEAN, Alistair. *Ice Station Zebera*. London, Collins, 1963. 255 pp.

Suspenseful novel re USS *Dolphin* (SSN) and a weather station at North Pole. Cold War setting.

822 McLEOD, Grover Stephen, 1923- . *Sub sailor*. Birmingham, Alabama, Manchester Press, 1964. 315 pp., illus.

Novel of WW II and USN submarines in the Pacific. Fictional submarine USS *Terrapin*. Author served in USN submarines during the period.

823 ________. *Teodoro*. Birmingham, Ala., Manchester Press, 1969. 254 pp., illus.

USN submarines in Pacific. Fiction. Phillipines guerilla operations.

MacNeil, James, *pseud*. See entry 813.

824 MADARIAGA, Juan de. *El Submarino Peral*. Madrid, 1889.

Re the Spanish submarine *Isaac Peral*.

Die Männer von U96. See entry 1459.

Magellan, Ferdinand, c.1480-1521. See entries 304, 305.

Maggie, of the Suicide Fleet. See entry 160.

825 MALKUS, Alida Sims, 1895- . *Exploring the Sky and Sea: Auguste and Jacques Piccard*. Chicago, Encyclopaedia Brittanica Press, 1961. Chicago, Kingston House, 1961. Illus. by Robert Boehmer. 190 pp., illus. LC 61-10686.

Juvenile work concerning the bathyscaphe Trieste.

826 MALLMAN-SHOWELL, J. P. *U-Boote Gegen England*. Stuttgart, Motorbuch Verlag, 1974. 200 pp., illus.

This is a German edition of a book published in England under the title *U-Boats Under the Swastika*. Contains very good pictures.

Malta submarines. See entry 1272.

Man and the Sea. See entry 496.

Man and the Underwater World. See entries 751, 752.

Man Beneath the Sea. See entry 958.

Man Explores the Sea. See entries 344, 1442.

Man-in-the-Sea. See entry 123.

Man-of-War: A History of the Combat Vessel. See entry 818.

Man Under the Sea. See entries 345, 346.

Man Under Water. See entry 107.

Manheim, Ralph. See entry 941.

Manned Submersibles and Underwater Surveying. See entry 1327.

Manoevering as a Defense Against Attack by Submarines. See entry 63.

Man's Rush to the Seafloor. See entry 404.

Mansergh, Aubrey. See entry 494.

A Manual for Free-Divers Using Compressed Air. See entry 945.

827 MANUS, Max, 1914- . *Underwater Saboteur.* Translated by F. H. Lyon. London, Kimber, 1953. 239 pp., plates.
Frogman and underwater demolition.

Marchand, René. See entry 709.

Marchegay, Gilberte. See entry 375.

Marcuse, Adolf. See entry 706.

Mare Island Navy Yard. See entry 798.

Margaret (submarine chaser). See entry 160.

828 MARINA, Victor. *Cartilla de Electricidad en su Aplicacion a las Defensas Submarinas para las Clases Subalternas de la Armada, Auxiliares en las Estaciones de Torpedoes.* Madrid, 1882. 200 leaves.

Marine Archaeology. See entries 116, 1219.

Marine Bio-Acoustics. See entry 1218.

La Marine Française Pendant la Grande Guerre (1914-1918). See entry 228.

Der Marine Nachrichten und Ortungsdienst. See entry 486.

Marine Salvage. See entry 497.

Marine Salvage Operations. See entry 138.

829 MARINE TECHNOLOGY SOCIETY. *Research Submersibles in Oceanography.* Washington, Marine Technology Society, 1970.

830 ________. UNDERSEA VEHICLE COMMITTEE. *Safety and Operational Guidelines for Undersea Vehicles.* Washington, Marine Technology Society, 1968. illus. LC 78-2203.
Manned underwater research stations.

Marine Technology Society Journal. See entry 691.

Marinekleinkampfmittel. See entry 410.

. . . *Maritime Danger Zones* . . . See entry 1316.

Maritime History of Russia, 848-1948. See entry 869.

Maritime Law. See entry 1383.

831 MARS, Alastair, 1915- . *Arctic Submarine.* London, Elek, 1955. 191 pp. LC 55-33530.

832 ________. *British Submarines at War: 1939-1945.* Annapolis, Naval Institute Press, 1974. 256 pp., illus.
Includes aspects of Allied submarine operations.

833 ________. *Fire In Anger.* NY, Mill, 1958. 222 pp. LC 58-7581.
Same with title *Atomic Submarines: A Story of Tomorrow.* London, Elek, 1957. Toronto, Ryerson, 1957. 192 pp. LC 58-18342.

834 ________. *HMS Thule Intercepts.* London, Elek, 1956. Toronto, Ryerson, 1956. 256 pp., plates, maps. LC 57-337.

835 ________. *Submarine at Bay.* London, Elek, 1956. Toronto, Ryerson, 1956. 164 pp., illus. LC 56-39513.

836 ________. *Unbroken: The Story of a Submarine.* London, Muller, 1953. Sydney, Shakespeare, 1953. Toronto, Saunders, 1953. 224 pp. LC 53-32575.

837 MARSCHALL, Wilhelm. *Torpedo Achtung! Los! Erleb-*

nisse im U-Bootkrieg, 1917/18. Berlin, Deutschen Verlag, 1938. 229 pp., illus., map, plates, ports. LC 39-8080.

Martin, Frederick. See entry 360.

838 MARX, Robert F. *They Dared the Deep: A History of Diving.* Cleveland and NY, World, 1967. 160 pp., illus., bibliog. LC 67-13830.

Good introductory survey from ancient times to present. Useful. Kind of sketchy on submarine history. Marx is an experienced diver and underwater archaeolgist. Worked on the excavation of Port Royal for the Jamaican government. Author of numerous books.

839 MASON, David, 1938- . *U-Boat, The Secret Menace.* NY, Ballantine, 1968. 160 pp., illus.

Pictorial record of German submarines their successes and eventual destruction.

840 ________. *U-Boat—The Secret Menace.* London?, Pan Books, 1972. 160 pp., illus., maps. ISBN 0-345-09732-7.

841 MASON, Francis van Wyck, 1897- . *Our Valiant Few.* Boston, Little Brown, 1956. Toronto, 1956. 436 pp. LC 56-10640.

Illustrated by John Alan Maxwell.

842 MASTERS, David, 1883- . *Deep-Sea Diving.* Illus. with photos and with drawings by L. R. Brightwell. London, NY, Nelson, 1935. 94 pp., frontis., port., illus., diagrams. LC 35-18782.

"Hard hat" diving. Salvage diving.

843 ________. *Divers in Deep Seas: More Romances of Salvage.* London, Eyre & Spottiswoode, 1938. 284 pp., frontis., illus. LC 39-9633.

844 ________. *Epics of Salvage: Wartime Feats of the Marine Salvage Men in WW II.* Boston, Little, Brown, 1954. 234 pp., illus. LC 54-8230.

Same. London, Cassell, 1953. 264 pp., illus. LC 53-33402.

845 ________. *I.D. New Tales of the Submarine War.* London, Eyre & Spottiswoode, 1935. 296 pp., frontis., plates, ports. LC 36-6931.

846 ________. *The Submarine War.* NY, Holt, 1935. xv, 296 pp., frontis., plates, ports. LC 35-23601.

American edition of entry 845.

847 ________. *Up Periscope.* NY, Dial, 1943. Toronto, Longmans, 1943. xxi, 275 pp. LC 43-10317.

Same. London, Eyre & Spottiswoode, 1942. Toronto, Collins, 1942. 176 pp., plates. LC 42-51954.

Same. Sydney, Austral-Asian Pub. co., 1945. 189 pp.

848 *The Mastery of Air, by the Author of The Triumph of Man.* London, NY, Pitman, 1922. 184 pp., frontis., illus., diagrams. LC 23-9430.

Matejko, Theo. See entry 713.

Mathematical Magick. See entry 1465.

849 MATKIN, Robert Broadbent and G. F. BROOKES. *Scuba Diver: The Path to Underwater Adventure.* London, MacDonald, 1963. 152 pp., illus.

Max Horton and the Western Approaches. See entry 207.

Maxim, Hudson. See entry 303.

850 MAYERS, Colin. *Submarines, Admirals and Navies.* Los Angeles, Associated Publications, 1940. 280 pp., frontis., plates, port. LC 40-29538.

Author is a retired Lieutenant Commander who was the senior instructional officer of the British submarine school.

Med Staalnerver. See entry 610.

Med Tyska U-Båtar och i Tyska Fångläger. See entry 222.

A Media Milla de Profundidad. See entry 83.

Medical Aspects of Sport Diving. See entry 343.

Mediterranean. See entry 1150.

851 MEIER, Frank. *Fathoms Below: Under-Sea Salvage from Sailing Ships to the Normandie.* NY, Dutton, 1943. 319 pp., frontis., plates, port. LC 43-4471.

852 ________. *Up for Air: Diving for a Living.* NY, Dutton, 1940. 287 pp., frontis., plates, ports. LC 40-27554.

Mein Wechselvolles Leben. See entry 317.

Mein Weg Nach Scapa Flow. See entry 999.

853 MEISTER, Juerg. *Der Seekrieg in de Osteuropaeischen*

Gewassern, 1941-1945. 392 pp., maps.

Mémoire sur la Navigation Sous-Marine. . . . See Entry 132.

Mémoire sur les Bateaux Sous-Marin. See entry 34.

Mémoire sur les Mines Flottantes. . . . See entry 878.

Mémoire sur une Nouvelle Machine à Plonger, Appelée Triton. See entry 340.

Memoirs: Ten Years and Twenty Days. See entry 318.

Memoria Acerco de Ictineo. . . . See entry 879.

Memoria Sobre La Navegacion. . . . See entry 880.

Memoria Sobre o Submarino Fontes. See entry 960.

Men Against the Sea. See entry 935.

Men in the News. See entry 908.

Men in the Sea. See entry 146.

Men Under the Sea. See entries 370, 768.

Men Under Water. See entry 348.

Merchant Marine. See entries 630, 1448.

Merchant raiders. See entry 1481.

Merchant Shipping Losses. . . . See entry 513.

Merchant Tonnage and the Submarine. See entries 514, 515.

854 MERRILL, James Mercer, 1920- . *The Rebel Shore: Story of Union Seapower in the Civil War.* Boston, Little, Brown, 1957. Toronto, 1957. 246 pp., illus. LC 57-11156.

Mersenne, Pere. See entry 295.

Mes Navires Mystérieux. See entry 182.

Met Hr. Ms. K XIII naar Nederlandsch-Indies. See entry 781.

855 METEL, Rasit. *Türk Denizalticilik Tarihi.* Istanbul, 1960. xiv, 959, 16 pp., illus, tables, facsims.

Comprehensive history of the Turkish submarine force written by a Turkish submarine officer.

856 METZLER, Jost. *The Laughing Cow: A U-Boat Captain's Story.* Translated by Mervyn Savill. London, Kimber, 1955.

217 pp., plates, maps.

Meutes Sous-Marines. See entry 166.

Mexico. See entry 170.

857 MICHELSEN, Andreas, 1869- . *Der U-Bootskrieg, 1914-1918.* Leipzig, Hase & Koehler, 1942. 207 pp., frontis., map, tables. LC 43-10409.

First published in 1925.

The Midget Raiders. See entry 1427.

Midget submarines. See entries 15, 51, 87, 88, 123, 129, 146, 153, 154, 396, 410, 466, 1019, 1257, 1425, 1427.

Midway. See entry 884.

858 MIELKE, O. *Deutschen U-Boote, 1939-1945 Ihr Verblieb, Tech. Daten, Typenliste, Namen der Kommandaten, etc.* Munich, 1959. 112 pp.

859 MILANESI, Guido, 1872- . *Sommergibili: Il Monge, H,3, U.C.12 i Nostri.* Milano, Alfieri & Lacroix, 1917. 64 pp., illus., ports. LC 18-11903.

860 ________. *Submarines: the Monge, the H.3, the U.C.12, Ours.* (Publications of the Special Department of the Ministry of Marine). Milan, Alfieri & Lacroix, 1916? 56 pp., illus., ports. LC 18-11904.

861 MILES, Stanley. *Underwater Medicine.* London, Staples Press, 1962. Phila., Lippincott, 1962. 328 pp., illus., bibliog. LC 62-14637.

Psychological approach to underwater environment, its hazards, its physiological effects. Data on water safety, submarine escape equipment, and marine animals. Author is a Royal Navy surgeon.

862 ________. *Underwater Medicine.* 2nd ed. London, Staples Press, 1966. (LC 67-76619) and Phila., Lippincott, 1966 (LC 66-8390). 332 pp., illus.

863 ________. *Underwater Medicine.* 3rd ed. London, Staples Press, 1969. 363 pp., illus. SBN 286-62723-X.

Milgram, N. See entry 725.

Miller, R. R. See entry 1388.

864 MILLET, J. B. *Submarine Signalling by Means of Sound.*

London, Institution of Naval Architects, 1905. 7 pp., folding plate. LC 6-13146.

Read at the Spring Meeting of the 46th session of the Institution of Naval Architects. April 14, 1905.

865 MILLHOLLAND, Ray. *The Splinter Fleet of the Otranto Barrage.* Indianapolis, Bobbs-Merrill, 1936. 307 pp., frontis., plates, ports., maps, diagram. LC 36-6968.

WW I mine warfare.

866 MILNE, Thomas Ferris. *This World and That, the Autobiography of a Diver.* Edited by Frank D. Long. Foreword by ADM Mark Kerr. London, Joseph, 1934. 234 pp., frontis., plates, ports., facsim. LC 34-35638.

867 MINARIK, William H. *Sailors, Subs, and Señoritas.* Boston, Brandon Press, 1968. 349 pp.

Memoirs of a navy submarine man.

Mine and Countermine. See entry 799.

Les Mines Sous-Marines. See entry 501.

Mines and Mining. See entries 4, 5, 6, 17, 91, 97, 114, 141, 211, 212, 264, 301, 322, 327, 332, 360, 501, 511, 520, 538, 555, 556, 558, 755, 761, 796, 799, 865, 878, 921, 937, 961, 1026, 1027, 1050, 1051, 1053, 1080, 1082, 1099, 1100, 1168, 1190, 1192, 1278, 1283, 1316, 1329, 1357, 1380, 1386, 1400, 1401.

Mines et Torpilles. See entry 1192.

Mines, Minelayers and Minelaying. See entry 264.

Minnesota Historical Society. See entry 245.

The Mirror of Alchemy. See entry 43.

Missile Base Beneath the Sea. See entry 306.

868 MITCHELL, Donald W. *A History of Russian and Soviet Sea Power.* NY, Macmillan, 1974. 657 pp., illus., maps, bibliog. LC 72-93629.

Includes photos of Russian submarines. Notes on sources. An exhaustive scholarly work.

869 MITCHELL, Mairin. *Maritime History of Russia, 848-1948.* London, Sidgwick, 1949. NY, Macmillan, 1949. xvi, 544 pp., plates, maps. LC 50-1526.

Many of the maps were specially drawn for this book by J. F.

Horrabin. Extensive bibliography between pp. 455-480. Submarine references scattered throughout.

870 MITTLER, Jost. *Die Deutsche Kriegsflotte und Ihre Verbündeten.* Berlin, E. S. Mittler, 1915-1922. 5 volumes.
Covers four years of sea warfare.

Moana (Vastness of the Waters). See entry 500.

Moana (cutter). See entry 499.

Model Submarines for Beginners. See entry 487.

Model Yachts, Sailing Boats and Submarines. See entry 586.

A Modern Submarine Boat. See entry 1222.

Modern Weapons of War, By Land, Sea, and Air. See entry 558.

Moderne Küsten U-Boote. See entry 922.

Das Moderne Unterseeboot. See entry 1415.

Los Modernos Barcos Submarinos. . . . See entry 877.

871 MÖLLER, Wilhelm. *Baldiger U-Boot-Friede?* Dresden, Mitteldeutsche Verlag, 1918. 71 pp., illus. LC 55-54524.

872 MOISEEV, S. P. *Spisok Korablei Russkogo Parovogo i Bronenosnogo Flota, s 1861 po 1917 g.* (List of vessels in the Russian steam and armored fleet from 1861 to 1917.) Moscow, 1948.

Molish, H. B. See entry 1353.

873 MOLLI, Giorgio. *La Vérité sur la Guerre Sous-Marine.* Preface by H. Henquinez. Brussels, Norz, 1917. 112 pp. LC 44-36320.

Momsen, Charles B. See entries 807, 1419.

874 MONASTEREV, Nestor, 1888- . *Histoire de la Marine Russe.* Paris, 1932.

875 ________. *Le Navire Sous-Marin.* Paris, Nouvelles Editions Latines, 1935. 90 pp.

Le Monde du Silence. See entry 259.

Monge (Italian submarine). See entries 859, 860.

876 MONOD, Théodore, 1902- . *Bathyfolages: Plongées Profondes, 1948-1954.* Paris, Julliard, 1954. 213 pp., illus. LC 54-33731.

Monsarrat, Nicholas. See entry 1091.

877 MONTERO y de TORRES, Enrique de. *Los Modernos Barcos Submarinos al Alcance de Todos.* Madrid, Orrier, 1912. 441 pp., illus., tables, map, diagrams, bibliog. LC 43-41009.

878 MONTGERY, J. P. *Memoire sur les Mines Flottantes et les Petards Flottants ou Machines Infernales.* Paris, 1819.

879 MONTURIOL, Narciso, 1819-1885. *Memoria acerca del Ictineo y Ensayo sobre el Arte de Navegar por Debajo del Agua.* Barcelona, 1891.

Monturiol was a mechanic and inventor who was born in Barcelona. He was in France in 1848, but left there because of his political ideas and activity. He built his submarine the *Ictineo,* and launched it in Barcelona, on June 28, 1859. Perfected it during 1860, and made successful demonstrations of its workability in Barcelona and in Alicante in 1861.

880 ________. *Memoria Sobre la Navegacion Submarina; por el Inventor del Ictineo o Barco Pez.* Barcelona, 1860.

881 [Monturiol] *Narciso Monturiol: Inventor del Primer Submarino. Biografia del Sabio Espanol y Explicaciones sobre su Ictineo.* Buenos Aires, Editorial Tor, 1916. 124 pp. LC 18-27448.

A rearranged and revised reprint of Estrany's work on Monturiol. See entry 382.

Moore, Don. See entry 389.

882 MORAHT, Robert. *Werewolf der Meere: U64 Jagt den Feind.* Berlin, Schlegel, 1934. 153 pp., illus., plates, ports., facsim. LC 34-23310.

WW I German U-boat.

Moran, Mabel (O'Connell) 1890- . See entry 172.

Der Mordversuch an Oberleutnant z.s. Crompton. See entry 110.

Morgan, William L. See entry 172.

883 MORI, K. *Submarine in War: A Study of Relevant Rules*

and Problems. NY, Stechert. London, Routledge.Tokyo, Maruzen. Toronto, Carswell, 1931. 185 pp. LC SD33-28.

884 MORISON, Samuel Eliot. *Coral Sea, Midway, and Submarine Actions.* (volume 4 of History of U.S. Naval Operations in WW II) Boston, Little, Brown, 1949. LC 47-1571.

Moro (Italian submarine). See entry 563.

885 MORRIS, Richard Knowles. *John P. Holland, 1841-1914: Inventor of the Modern Submarine.* Annapolis. U.S. Naval Institute, 1966. 211 pp., illus., diagrams, bibliog. LC 66-20239.

886 MORRISS, Frank. *Submarine Pioneer, John Philip Holland.* Illus. by Dirk Gringhuis. Milwaukee, Bruce, 1961. 144 pp., illus. LC 61-14750.

Morton, Dudley "Mush." See entry 1185.

The Most Formidable Thing. See entry 672.

Motivation. See entry 1353.

De Motu Animalum. See entry 127.

887 MUNDY, Arthur Joseph. 1850-1912. *Acoustic Traingulation and its Application to Submarine Signalling.* Boston, Millet, 1900. 16 pp., diagrams. LC 0-3919 Revised.

Munitions Industry. See entry 1314.

Murder at Sea. See entry 645.

Murphree, Henry B. See entry 1344.

888 MUSEUM OF SCIENCE AND INDUSTRY. *The Story of the U-505.* Chicago, Museum of Science and Industry, 1955. 36 pp., illus.

Has a foreword by ADM Daniel V. Gallery, man who was in charge of the task force which took U-505 as a prize during WW II.

Muskallunge, USS (SS 262). See entry 1318.

My Love Affair with the Navy. See entry 131.

My Mystery Ships. See entries 183, 184.

My War at Sea. See entry 1111.

889 MYERS, John James, 1920- and others, editors. *Hand-*

book of Ocean and Underwater Engineering. (Prepared under the auspices of the North American Rockwell Corp.) NY, McGraw Hill, 1969. 1070 pp., illus. SBN 07-044245-2. LC 67-27280.

N

890 NABER, H. A. *De Ster van 1572: Cornelis Jacobsz Drebbel, 1572-1633.* Amsterdam. Wereld, 1907. 92 pp., illus.

891 NADAUD, Marcel. *Les Patrouilleurs de la Mer: Souvenirs de Guerre Sous-Marine.* Preface by Charles Le Goffic. Paris, Michel, 1918? 184 pp., plates, port. LC 19-16549.

The Naked Warriors. See entry 389.

892 NALL, Jane. *Submarine Habilability: A Literature Survey (1951-1959).* Washington, Documentation, Inc., 1959. 31 pp.

Narciso Monturiol. . . . See entry 881.

Nastavlenie po Vodolaznemu Delu. See entry 1083.

893 NATIONAL ACADEMY OF SCIENCE. COMMITTEE ON UNDERSEA WARFARE. *Status of Research in Underwater Physiology.* Washington, NAS, 1956.

National Defense Research Committee. See entry 595.

894 NATIONAL RESEARCH COUNCIL. COMMITTEE ON UNDERSEA WARFARE. *An Annotated Bibliography of Submarine Technical Literature, 1557 to 1953.* (National Research Council. Publication 307) Washington, 1954. xiii, 261 pp. LC 54-60800.

895 ________. ________. *A Survey Report on Human Factors in Underseas Warfare.* Washington, National Re-Search Council, 1949. 541 pp., diagrams, tables.

________. Supplement. 1954. 92 pp., diagrams, tables.

896 ________. *Mine Advisory Committee. Proceedings of the Second Symposium on Underwater Physiology.* February 25-26, 1963. Washington. Edited by Christian J. Lambertson and Leon J. Greenbaum, Jr. Prepared for the Office of Naval Research by the Committee. 1963. Illus.

De Natura Elementorum. See entry 338.

Nautical Archaeology. See entry 1462.

Nautilus (French submarine). See entry 764.

Nautilus (Fulton's). See entries 237, 453-457, 951.

Nautilus (Norwegian submarine). See entry 1206.

897 THE NAUTILUS. *Pictorial and Historical Events of the World's First Nuclear-Powered Vessel, the USS Nautilus (SSN 571).* Atlanta, Albert Love, 1958. 221 pp., illus.

Prepared and edited by Nautilus crewmen. Covers events of 1951-1957. Cruise book of the Nautilus' career under CDR Wilkinson, before the Arctic transit.

Nautilus, USS (SSN 571). See entries 18-23, 117, 118, 633, 897, 1095, 1438.

Nautilus (Wilkins). See entry 1463.

Du Nautilus au Redoutable. See entry 764.

Nautilus 90 Grad Nord. See entry 1095.

Nautilus 90° Nord. See entry 21.

The Naval Battles of the Russo-Japanese War. See entry 1252.

898 NAVAL HISTORY DIVISION. *The Submarine in the United States Navy.* Washington, Naval History Division, 1960. 32 pp., illus., diagrams.

899 ________. *The Submarine in the United States Navy.* Washington, 1963. Foreword by Chester W. Nimitz. 23 pp., illus., diagrams.

A Naval History of the War, 1914-18. See entry 909.

Naval Inventions. See entry 1144.

Naval Mining and Degaussing. See entry 796.

The Naval Service of Canada. See entry 1271.

The Naval War in the West: The Wolf Packs. See entry 354.

La Nave Sommergibili. See entry 542.

La Nave Subacquea. See entry 181.

Navies of the Second World War: series. See entries 770-773.

Navigation Aérienne et Navigation Sous-Marine. See entry 919.

Navigation Sous-Marine. See entry 302.

La Navigation Sous-Marine. See entries 465, 964, 965, 1008, 1414.

La Navigation Sous-Marine à Travers les Siècles. See entry 299.

Navios Perdidos. See entry 438.

Le Navire Sous-Marin. See entry 875.

The Navy Bluebook. See entry 243.

Navy Diver. See entry 398, 695.

The Navy Hunts the CGR 3070. See entry 1238.

De Nederlandsche Koopvaardij. . . . See entry 624.

Netherlands. See entry 1302.

900 NETHERLANDS. Dept. van Buitenlandsche Zaken. *Diplomatieke Bescheiden Betreffende den Verscherpten Duikbootoorlog.* 's Gravenhage, Algemeene Landsdrukkerij, 1917. 4 pp. LC 18-15177.

Correspondence between the Netherlands foreign ministry and the ambassadors from Germany and Austria-Hungary.

Netherlands Marine. See entry 241.

Netherlands Merchant Marine. See entry 624.

Netherlands submarines. See entry 781.

923 Meter Unter dem Meeresspiegel. See entry 84.

901 NEURATH, Marie Reidemeister. *Exploring Under the Sea.* NY, Lothrop, 1960. 36 pp., illus. LC 60-16061.

Juvenile.

902 NEUREUTHER, Karl, ed. *U-Boat Stories: Narratives of German U-Boat Sailors.* Edited by Karl Neureuther and Claus Bergen, with illus. by Claus Bergen. Translated by Eric Sutton. . . . NY, Long and Smith, 1931. London, Constable, 1931. 207 pp., frontis, illus., col. plates. LC 32-2479.

903 ________. *Das Unterseeboot.* Munich, Hertz, 1915. 23 pp., diagram.

904 ________. *Wir Leben Noch! Deutsche Seehelden im U-Bootkampf. 26 Erlebnisse von Amgehörigen der U-Boot-kameradschaft Munchen.* with Claus Bergen. Stuttgart, Union Deutsche Verlagsgesellschaft, 1930. 206 pp., frontis., illus., col. plates. LC 31-5052.

905 ________. *Wir Leben Noch! Deutsche Seehelden im U-Boot-Kampf.* Stuttgart, Union, 1941. 206 pp., 8 color plates, illus.

Many of the illus. are by Claus Bergen.

Neutralitat, Blockade und U-Bootkrieg. . . . See entry 159.

906 NEVSKY, N. A. *Voenno-Morskoi Flot.* Moscow, 1959.

New Submersible Battle Cruiser with Sixteen Inch Gun. See entry 147.

907 NEW YORK. PUBLIC LIBRARY. *Submarines: A List of References in the New York Public Library.* Compiled by Mary Ethel Jameson. Science Division, with a foreword by Simon Lake. NY, 1918. 97 pp. LC 18-9668.

908 NEW YORK TIMES. *Men in the News, 1958-1959.* vols. 1 & 2. Edited by Robert H. Phelps. Phila. Lippincott, 1959, 1960. Toronto, Longmans, 1959, 1960. LC 59-7787.

James Calvert is referred to in vol. I, p. 50.

New Zealand Navy. See entry 24.

909 NEWBOLT, Sir Henry John, 1862-1938. *A Naval History of the War, 1914-18.* London, Hodder, 1920. 350 pp., illus., maps. LC 21-1389.

910 ________. *A Note on the History of Submarine War.* NY, Doran, 1917? 26 pp. LC 18-22623.

911 ________. *Submarine and Anti-Submarine.* Illustrations

by Norman Wilkinson. London, NY, Longmans, Green, 1918. 312 pp., illus., plates. LC 19-1601.

912 NEWCOMB, Richard F. *Abandon Ship! Death of the USS Indianapolis.* NY, Holt, 1958. Toronto, McLeod, 1958. 305 pp., illus. LC 58-11218.

Newnes Complete Guide to Underwater Swimming. See entry 993.

913 NEWPORT MANUFACTURING CO. *The Invisible Destroyer.* Newport, R.I., 1902? 41 pp., illus.
Running title: "The Story of a Submarine Boat."

Newspapers. See 1195.

914 NEWTON, R. N. *Practical Construction of Warships.* London, Longmans, 1966. 464 pp., diagrams.
An excellent source book about construction of submarines.

Nicosian (ship). See entries 477, 479.

915 NIEMÖLLER, Martin, 1892- . *From U-Boat to Pulpit.* London, Hodge, 1936. 217 pp., frontis., plates, port. LC 37-23490.
Translated by CDR. D. Hastie Smith.

916 ________. *From U-Boat to Pulpit* with an appendix *From Pulpit to Prison* by Henry Smith Leiper. Chicago, NY, Willett, Clark & co., 1937. 233 pp., ports. LC 38-27058.

917 ________. *From U-Boat to Concentration Camp: The Autobiography of Martin Niemöller* with his further story by the Dean of Chichester. London, Hodge, 1939. 281 pp., ports. LC 39-15045.

918 ________. *Vom U-Boot zur Kanzel.* Berlin, Warneck, 1934. 210 pp., illus., plates, ports, map. LC 35-6580.

Night of Terror. See entry 203.

Night Raider of the Atlantic. See entries 1045, 1046.

Nimitz, Chester W. See entries 790, 791, 792, 899.

1917, Februar, 1918. . . . See entry 775.

Ninth Time Lucky. See entry 1257.

Nitsche, Erik. See entry 473.

Niven, John. See entry 473.

919 NOALHAT, Henri. *Navigation Aérienne et Navigation Sous-Marine: Deux Faces d'un Même Probleme.* Paris, Geisler, 1910. 200 pp., illus., diagrams. LC 34-31491.

920 ________. *Les Sous-Marins et la Prochaine Guerre Navale.* Paris, Berger-Levrault, 1903. 246 pp., plate, table, illus.

921 ________. *Les Torpilles et les Mines Sous-Marines.* Preface by Paul Fontin. Paris, Nancy, Berger-Levrault, 1905. lix, 432 pp., illus., ports, folding plate, diagrams. LC 25-14003.

Noalhat, Henri. See also entry 414.

922 NOHSE, Lutz and Eberhard ROESSLER. *Moderne Küsten U-Boote.* Munich, Lehmann, 1972. 140 pp., illus., diagrams.

Technical treatise on the development of coastal submarines.

No Less Renowned. See entry 551.

Noises. See entry 1368.

923 NOLI, Jean, 1928- . *The Admiral's Wolf Pack.* Translated by J. F. Bernard. NY, Doubleday, 1974. 396 pp. ISBN 0-385-00372-2. LC 72-89338.

Originally published in France in 1970 with title *Les Loups de L'Amiral.* Story of WW II U-boat warfare under Admiral Doenitz. Author interviewed U-Boat veterans, and had access to the Kriegsmarine combat records.

None More Courageous. See entry 618.

Norris, Paul. See entry 1407.

924 NORSK VAREKRIGSFORSIKRING A/S. *Varekrig: Utgitt av Styret for A/s Norsk Varekriegsforsikring, Stiftet 8, September 1914.* Oslo, Fabritius, 1931. 640 pp., tables, diagrams, plates, ports, formulae. LC 32-6694.

Economic aspects of WW I. War risk insurance. Effect of U-boat war on Norwegian shipping.

Det Norske Ubåtvåpen; 1909-1959. See entry 1068.

The Northern Barrage and Other Mining Activities. See entry 1386.

Norway. See entries 944, 1461.

925 NORWAY. SJØFARTSKONTORET. *Pertes de Guerre Subies par la Marine de Commerce Norvégienne 1914-1918.* . . . Kritiana, Kølbelske, 1918? 117 pp., tables. LC 21-3956.

Effects of the U-boat war on Norwegian shipping and the economy of Norway.

926 ________. ________. *Rapports de Mer Sur les Pertes de Guerre Subies par la Marine de Commerce Norvégienne.* . . . Kristiania, Centraltrykkeriet, 1917-19. 4 vols., table. LC21-3957.

927 ________. ________. *Sjøforklaringer Over Norske Skibes Krigsforlis.* . . . Kristiana, Centraltrykkeriet, 1917-18. 5 vols., tables. LC 21-3955.

Norway, Nevil Shute, 1899-1960. See Shute, Nevil.

Norwegian Merchant Marine. See entries 925, 926, 927.

Norwegian Submarines. See entries 1068, 1206.

I Nostri Sommergibili Durante la Guerra 1915-18. See entry 667.

A Note on the History of Submarine War. See entry 910.

Notes on Submarine Mining in England. . . . See entry 5.

Notice sur Denis Papin. See entry 47.

Novum Organum Scientarium. See entry 40.

928 NOYES, Alfred, 1880- . *Jakten pa U-Batar.* Stockholm, Lundberg & Olzon, 1916. 79 pp. LC 25-19619.

928a ________. *Mystery Ships: Trapping the U Boat.* London, Hodder, 1916. 181 pp.

929 ________. *Open Boats.* NY, Stokes, 1917. 91 pp., illus. LC 17-13295.

Nuclear Navy, 1946-1962. See entry 606.

930 *Nuclear Powered Submarines.* Foreword by H. G. Rick-

over. Oak Ridge, Tenn., U.S. Atomic Energy Commission Division of Technical Information, ca. 1965. 56 pp., illus.. diagrams, bibliog.

Nuclear-Powered Submarines, 1950-1965. See entry 122.

Nuclear Submarine Skippers and What They Do. See entry 1175.

Nuovo Trattato . . . Con Una Dissertazione sul Modo di Navigare Sott'acqua. . . . See entry 1254.

Nuremburg Prison. See entry 698.

931 NUTTING, William Washburn. *The Cinderellas of the Fleet.* With stories and photographs by Walter P. Groszman, Herbert L. Stone, Joseph L. Day, and others. Frontispiece by John Olaf Todahl. Jersey City, N.J., The Standard Motor Construction Co., 1920. 178 pp., illus. LC 20-9394.

About submarine chasers.

932 NYON. CONFERENCE, 1937. *International Agreement for Collective Measures Against Piratical Attacks in the Mediterranean by Submarines.* Nyon, September 14, 1937. London, HMSO, (Gt. Brit. Foreign Office. Treaty Series, no.38. 1937.) 16 pp., folding map. LC 40-749.

Deals with submarine operations in the Mediterranean during the Spanish Civil War of 1936-1939.

O

The Ocean Adventure. See entry 1153.

Ocean Engineering. See entry 1225.

The Ocean Laboratory. See entry 1164.

Ocean Vehicles. See entry 8.

Oceanography. See entries 220, 225, 233, 250, 265, 657, 829, 889, 1207, 1233, 1246, 1327, 1442.

Oceanology. See entry 958.

Ocherki po Bor'be s Podvodnymi Lodkami. See entry 1262.

933 O'CONNOR, Raymond, comp. *The Japanese Navy in World War Two*. Annapolis, Naval Institute Press, 1969. 192 pp., illus.

Odum, T. See entry 1393.

934 OFFICE OF NAVAL INTELLIGENCE. *Post Mortems on Enemy Submarines*. Washington, 1942-45. Serials 1-16, ONI 250-G, 1942-1945.

O'HARA, G. J. See entry 1389.

935 OLNEY, Ross Robert, 1929- . *Men Against the Sea*. NY, Grosset & Dunlap, 1969. 147 pp., illus. ISBN 0-448-04456-0. LC 68-29978.

Underwater exploration for a juvenile audience.

936 OLSEN, Robert I. and David PORTER. *Torpedoes Away!* NY, Dodd Mead, 1957. 247 pp. LC 57-7446.

Fiction re WW II submarine operations.

On Board a United States Submarine. See entry 1227.

On The Beach. See entry 1131.

On The Bottom. See entries 371, 372.

On Using Self Contained Underwater Breathing Apparatus. See entry 557.

One Man Band. See entries 155, 156.

One of our Submarines. See entry 1489.

130 Feet Down, Handbook for Hydronauts. See entry 443.

O'NEAL, H. A. See entry 1393.

Only Four Escaped. See entry 1428.

937 *Opasni Prostori Zbog Mina u Jadranskom i Jonskom Moru.* Split, Yugoslavia, 1952- . Biennial. Issued by the Hidrografski Institut. LC 55-33464.

Bulletins re mines.

Open Boats See entry 929.

Open Fire! See entry 639.

Open the Ports: The Story of Human Minesweepers. See entry 538.

Operation Barney. See entry 790.

Operation Hideout. See entry 1340.

938 ORAM, H. P. K. *Ready for Sea.* London, Seeley Service, 1973. 245 pp.

Memoirs of a seaman, including wartime service in British submarines.

939 ORBELI, Ruben Abgarovich, d. 1943. *Issledovaniia i Izyskaniia.* Moscow, 1947. 282 pp., illus., port, map, facsims. LC 49-33723.

Ordeal by Sea. See entry 646.

Ordeal by Water. See entry 697.

Ordnance. See entry 1358.

The Organization of American Women for Strict Neutrality. See entry 1284.

940 ORTZEN, Len. *Stories of Famous Submarines.* London, Barker, 1973. 170 pp. ISBN 0-213-16422-1. LC 73-165839.

Orzel's Patrol. See entry 1151.

Otranto Barrage. See entry 865.

941 OTT, Wolfgang, 1923- . *Sharks and Little Fish.* Translated by Ralph Manheim. NY, Pantheon, 1958. Toronto, McClelland, 1958. 431 pp. LC 57-10236.

Documentary fiction of WW II U-boat service by a man who served in Atlantic war zone.

942 OTTO, Freidrich, 1887- . *Das Unterseeboot im Kampfe.* Leipzig, Amelang, 1915. 157 pp., frontis., illus. LC 15-15184.

Otto Weddigen: Ein Lebensbild. See entry 1034.

Otto Weddigen und Seine Waffe. . . . See entry 1441.

Otto Weddigen und Seine Waffe. See entry 705.

Our Valiant Few. See entry 841.

Out of the Mists. See entry 194.

Outlawry at Sea. See entry 647.

943 OUVAROFF, Serge, 1913- . *Torpilles Humaines.* Paris, Sfelt, 1951. 201 pp. LC 52-29480.
Underwater demolition teams. Frogmen.

944 ØVRESETH, I. *Vi Som Var Våbenløse: En Skibsførers Erindringer Fra Krigstiden 1914-18.* Oslo, Aschehoug, 1932. 242 pp., illus., plates, maps. LC 33-1193.

Overdue and Presumed Lost. See entries 1126, 1127.

945 OWEN, David M. *A Manual for Free-Divers Using Compressed Air.* London, NY, Pergamon Press, 1955. 62 pp., illus. LC 55-2856.

946 OWEN, John. *The Shadow in the Sea.* NY, Dutton, 1972.
Novel re British secret service agents versus a Russian submarine.

P

Paddle, USS (SS 263). See entry 1318.

Padgett, Lemuel Phillips, 1855- . See entry 1287.

947 PAINE, Ralph Delahaye. *The Fighting Fleets.* Boston, Houghton Mifflin, 1918. 392 pp. LC 18-12608.

948 PAINE, Thomas O. *Submarining: Three Thousand Books and Articles.* (TEMPO Center for Advanced Studies) Santa Barbara, California, General Electric Co., 1971. 414 pp. LC 75-121699.

Separate listings of works by author and by title. Includes periodical articles and report literature. Some annotations. Locates copies in Libraries and private collections. Author served in USN submarines during WW II.

949 PAPIN, Denis, 1647-1712. *Receuil de Diverses Pieces Touchant Quelques Nouvelles Machines.* Paris?, 1695.

Papin, Denis, 1647-1712. French physicist for a time associated with Boyle in England. During period of 1687-96 he held the chair of mathematics at University of Marburg. His theoretical work led to the development of a steam engine. See also entries 47, 381, 1089, 1484.

De Papino et Machina Papiniana. See entry 1484.

Parker, Gene. See entry 197.

950 PARRISH, Thomas. *Victory at Sea: The Submarine.* NY, Ridge Press for Scholastic Book Service, 1959. 60 pp. LC 60-1079.

Role of the submarine during World War Two.

Parsons, Charles. See entry 661.

951 PARSONS, William Barclay, 1859-1932. *Robert Fulton and the Submarine.* NY, Columbia University Press, 1922. xiii, 154 pp., frontis., port, plates, facsims. LC 23-5456.

Same. Reprinted by AMS Press. ISBN 0-404-04888-9.

Partridge, Benjamin Waring. See entry 220.

Les Patrouilleurs de la Mer. See entry 891.

952 PAULI, D. C. and G. P. CLAPPER. *Project Sealab Report. Experimental 45-Day Undersea Saturation Dive at 205 feet.* Sealab II Project Group. With lists of references. Edited by D. C. Pauli and G. P. Clapper. March 8, 1967. Naval Research Office. Defense Department. ACR series. D210-15: ACR-124. vi, 428 pp., illus.

953 PAVLOVICH, N. B. *Flot v Pervoi Mirovoi Voine.* (The Navy in the First World War). Moscow, 1964.

Payzant, Charles. See entries 1117, 1118, 1119.

Peace Moves and U-Boat Warfare. See entry 109.

Peace or War? See entry 1284.

Pearl Diver. See entry 100.

Pearl Diver's Luck. See entry 93.

Pearson, Michael. See entry 1193.

954 PECHMANN, W. B. von. *U-Bootsarbeit und ihr Erfolg.* Foreword by Kapitanleutnant v. Koblinski. Wien, Phoebus, 1918. 19 pp.

Peckelsheim, Spiegel. See Spiegel.

955 PEILLARD, Léonce, 1898- . *Geschichte des U-Bootkrieges 1939-45.* Vienna, Neff, 1970. 471 pp., illus.

956 ________. *The Laconia Affair.* Translated by Oliver Coburn. NY, Putnam, 1963. Don Mills, Longmans, 1963. 270 pp., plates. LC 63-9669.

Published in England by Jonathan Cape under title *U-Boats to the Rescue: The Laconia Incident.*

Peirano, J. H. See entry 1346.

Peligro es mi Profesión. See entry 269.

Pellini, W. S. See entry 1391.

957 PENFIELD, Thomas, 1903- . *U.S. Submarines in Action, a Fact Story of U.S. Submarines in the Pacific.* Racine, Wis., Whitman, 1944. 24 pp., col. illus. LC 44-39543.

Juvenile.

958 PENZIAS, Walter and M. W. GOODMAN. *Man Beneath the Sea.* NY, Wiley-Interscience, 1973. 831 pp., illus.

Comprehensive review of underwater engineering: past, present and future.

959 PERAL y CABALLERO, Antonio Isaac, 1851-1895. *El Profundo Isaac.* Madrid, 1934.

Peral y Caballero, Antonio Isaac, 1851-1895. See entries 962, 1412.

Peral, Marino de España. See entry 1412.

960 PEREIRA de MELLO, Joao Augusto Fontes, 1780-1856. *Memoria sobre o Submarino Fontes.* Lisboa, Typographia Industrial Portugueza, 1902. 70 pp., plates.

961 PERLIA, Zigmund Naumovich. *Udar Pod Vodoi.* Moscow, 1945. 246 pp., illus. LC 52-40214.

Torpedoes and mines.

962 PÉREZ, Dionisio, 1871- . *Isaac Peral: La Tragedia del*

Submarino Peral. Madrid, Ediciones Nuestra Raza, 1935. 186 pp. LC 36-4512.

Il Pericolo Sottomarino. See entry 682.

Pericot, Luis. See entries 1032, 1033.

Periscope Patrol. See entry 1272.

Der Perlentaucher. See entry 101.

Perles, Requins et Coraux. See entry 422.

963 PERRY, Milton F. *Infernal Machines: The Story of Confederate Submarine and Mine Warfare.* Baton Rouge, Louisiana State University Press, 1965. 231 pp., illus., appendices. LC 65-12838.

Perseverance Island, or The Robinson Crusoe of the 19th Century. See entry 437.

Pertes de Guerre Subies par la Marine de Commerce Norvégienne 1914-1918. See entry 925.

964 PESCE, G. L. *La Navigation Sous-Marine.* Paris, Becus, 1897. 147 pp., illus.

965 ________. *La Navigation Sous-Marine.* Paris, Vuibert et Nony, 1906. 498 pp., illus.

"One of the best historical studies on the subject, beginning with the diving apparatus, whether boats or bells, referred to in the Greek and Latin classics. There are reproductions of many rare prints and illustrations of boats long forgotten."—NAS/NRC bibliog.

966 PETERSON, Mendel Lazear, 1918- . *History Under the Sea: A Handbook for Underwater Exploration.* Washington, Smithsonian Institution, 1965. 108 pp., illus., plates. LC 65-61491.

967 ________. *History Under the Sea: A Handbook for Underwater Exploration.* Washington, Smithsonian Institution Press, 1969. 208 pp., illus. LC 68-9580.

968 PETROV, M. A. *Podgotovka Rossii k Mirovoi Voine po More.* (Russia's Preparedness at Sea for the World War.) Moscow, 1926.

Philippines. See entries 314, 661, 823.

Phillips, Dean. See entry 609.

Philmus, Lois C. See entry 178.

Photography, Underwater. See entries 78, 227, 259, 260, 347, 353, 422, 581, 597.

Physiologie de la Plongée. See entry 300.

The Physiology and Medicine of Diving and Compressed Air Work. See entry 95.

Physics of Sound in the Sea. See entry 102.

Piatt, V. R. See entries 1388, 1392.

969 PICCARD, Auguste, 1884-1962. *Earth, Sky and Sea.* Translated by Christina Stead. NY, Oxford, 1956. 192 pp., illus. LC 56-9276.

970 ________. *In Balloon and Bathyscaphe.* Translated by Christina Stead. London, Cassell, 1956. Toronto, British Book Service, 1956. 192 pp., plates. LC 56-32358.

Piccard, Auguste, 1884-1962. See also entries 402, 622, 623, 825, 1419.

971 PICCARD, Jacques, 1922- and R. S. DIETZ. *Seven Miles Down: The Story of the Bathyscaph Trieste.* NY, Putnam, 1961. Toronto, Longmans, 1961. 249 pp., plates. LC 60-16679.

972 PICCARD, Jacques, 1922- . *The Sun Beneath the Sea.* Translated by Denver Lindley. NY, Scribners, 1971. 405 pp., illus. SBN 684-31101-1. LC 76-123854.

The cruise of the Ben Franklin research submersible in the Gulf Stream.

Piccard, Jacques, 1922- . See also entries 825, 1419.

Piccard, Jean Felix, 1884-1963. See entries 622, 623.

Pictorial and Historical Events of the World's First Nuclear Powered Vessel, the USS Nautilus. See entry 897.

A Pictorial History of Oceanographic Submersibles. See entries 1207, 1208.

Pigboats. See entries 369, 373, 1069.

Pini, Wladimiro, 1879- . See entry 1166.

973 PINNEY, Roy, 1911- . *Underwater Archaeology: Treasures*

Beneath the Sea. NY, Hawthorne, 1970. 214 pp., illus. LC 77-122245.

Juvenile.

Les Pionniers de la Plongée. See entry 244.

Pirate Submarine. See entry 1446.

The Pirate's Progress. See entry 30.

974 PITERSKY, Nikolai Alekseevich, ed. *Boevoi put Sovetskogo Voenno-Morskogo Flota.* (Battle Path of the Soviet Navy). Moscow, 1964.

Also a German edition: *Die Sowjet-Flotte im Zweiten Weltkrieg.* Oldenburg, 1966.

975 PITMAN, Sir Isaac & Sons. *Submarine X7, and Other Tales in Pitman Shorthand.* London, Pitman, ca. 1918.

976 PITT, Barrie. *Zeebrugge-St. George's Day, 1918.* London, Cassell, 1958. 237 pp., plates. maps.

977 P.I.X. *The Spider Web, the Romance of a Flying-Boat War Flight.* Edinburgh and London, Blackwood, 1919. 278 pp., frontis., plates, maps, plan. LC 20-92.

Anti-submarine patrol.

978 PLATH, Karl. *U-Kreuzer 151 Greift An: U-Kreuzerfahrten nach Afrika und Amerika.* Potsdam, Voggenreiter, 1937. 128 pp., illus.

La Plongée. See entry 426.

Plongées sans Câble. See entry 1214.

Le Plongeur de Bourgois et Brun. See entry 133.

979 POCOCKE, Richard. *Travels in England.* London?, 1754.

Account of diving operations.

Podgotovka Rossi k Mirovoi Voine po More. See entry 968.

Podmornica. See entry 1086.

Podvodniki. See entry 1492.

Podvodnye Lodki v Operatsiiakh Russkogo Flota no Baltiskom More v 1914-1915 g g. See entry 1253.

Podvodnye Lodki v Russkom i Sovetskom Flote. See entry 1269.

Podvodnye Mastera. See entry 1495.

Poetry. See entry 1273.

980 POHL, Frederick and Jack WILLIAMSON. *Undersea City.* NY, Gnome Press, 1958. 188 pp. LC 57-14673.

Science fiction.

981 ————. ———— *Undersea City.* London?, Dobson, 1968. 188 pp. SBN 234-77222-0.

982 ————. ————. *Undersea Fleet.* NY, Gnome Press, 1956. 187 pp. LC 55-12189.

983 ————. ———— *Undersea Quest.* NY, Gnome Press, 1954. 189 pp. LC 54-7526.

984 POHL, Heinrich, 1883- . *Der Deutsche Unterseebootkrieg: Ein Vortrag Gehalten am 17. Februar 1925 auf dem Stuttgarter Juristenabend.* Stuttgart, Enke, 1925. 54 pp. LC A 31-236.

International law and the submarine.

Polaris. See entries 38, 306, 651, 1035, and 1279.

Polaris! See entry 38.

985 *Polaris Chronology. History of the Fleet Ballistic Weapon System Development Program, 1955-1963.* 12 pp.

No place or publisher indicated. Appears to be a lobbyist's "give-away" for Congress and the press.

986 *Polaris Fleet Ballistic Missile Weapon System. Fact Sheet 1 June 1966.* 40 pp., illus., tables.

No place or publisher indicated. Apparently a lobbyist's "give-away" for Congress and the press.

The Polaris Missile Strike. See entry 732.

Polish submarines. See entry 1151.

Politische und Militärische Bedeutung des Unterseebootskrieges 1914/18. . . . See entry 536.

Polk, Frank Lyon. See entry 448.

987 POLLINA, Paolo M., comp. *I Sommergibili Italiani 1895-1962.* Revised by Aldo Cocchia. Rome, Ufficio Storico Marina Militare, 1963. 300 pp., illus., diagrams.

988 POLMAR, Norman. *Atomic Submarines.* NY, Van Nos-

trand, 1963. 286 pp., illus. LC 63-5029.

History and development of the USN nuclear submarine fleet.

989 ________. *Death of the Thresher.* Phila., Chilton, 1964. Toronto, Ambassador, 1964. 148 pp., illus. LC 64-15134.

Loss of USN atomic submarine *Thresher* (SSN 593).

990 ________. *Soviet Naval Power.* Washington, National Strategic Information Center, 1973.

991 POOLE, Lynn. *Diving for Science.* Pictures by Jeanne Bendick. NY, Whittlesey House, 1955. 100 pp., illus. LC 55-9550.

Skin and scuba diving, scientific research. Marine biology.

Porter, David. See entry 936.

Les Portes de la Mer. See entries 310, 311.

992 PORTSMOUTH NAVAL SHIPYARD. *From Sails to Atoms.* Portsmouth, N.H., Naval Shipyard, 1963. 36 pp., illus.

Portsmouth Navy Yard. See entry 149.

Portsmouth, New Hampshire. See entry 149.

The Position of Enemy Merchantmen. See entry 1194.

Post Mortems on Enemy Submarines. See entry 934.

The Postwar Rearmament of Japanese Maritime Forces 1945-71. See entry 35.

993 POULET, Guy, 1924- and Robert BARINCOU. *Newnes Complete Guide to Underwater Swimming.* Translated by Douglas Burton. English technical adviser Mike Busuttill. London, Newnes, 1964. 310 pp., illus., plates, some in color.

994 POULEUR, Hector. *Torpilleurs et Sous-Marins. La Tactique Navale et les Enseignements de la Guerre Russo-Japonaise.* Liege, Desoer, 1904. 74 pp., plates.

Pour en Finir Avec les Sous-Marins. See entry 297.

Pownall, C. A. See entry 794.

Practical Construction of Warships. See entry 914.

Pratt, William V. See entry 800.

995 PRAVDIN, A. A. *Konstruktsiia Podvodnykh Lodok.* Moscow, 1947. 282 pp., diagrams. LC 49-20813.

Construction of submarine boats.

Prendergast, Maurice. See entry 484.

Prepare to Dive! The Story of Man Undersea. See entry 232.

. . . *Pressure-Vessel Materials.* See entry 1391.

996 PRESTON, Antony, ed. *Warships in Profile: Volume 3.* NY, Doubleday, 1974. 152 pp., illus.

One of the six chapters concerns "*Rubis,* Free French submarine, 1932-1957."

997 PRICE, Alfred. *Aircraft Versus Submarine.* London, Kimber, 1973. 268 pp., illus.

Anti-submarine warfare, employment of aircraft from 1912 to the present.

998 PRIEN, Gunther, 1908-1941. *I Sank the Royal Oak.* Translated by the Comte de la Vatine. London, Grays Inn Press, 1954. 196 pp., illus., ports., maps.

German submarine commander Prien took his U-47 into Scapa Flow and sank HMS Royal Oak.

999 ________. *Mein Weg Nach Scapa Flow.* Berlin, Deutscher Verlag, 1940. 190 pp., illus.

Prien, Gunther, 1908-1941. See also entries 430, 433, 819.

El Primer Viaje del Submarino Mercante Deutschland. See entry 716.

Prisoner of the U-90. See entry 664.

Proceedings of the Second Symposium on Underwater Physiology. See entry 896.

El Profundo Isaac. See entry 959.

Program for Uddannelsen af Mandskabet . . . Sømineafdeling. See entry 301.

Project Sealab. See entry 1117.

1000 *Project Sealab Report. An Experimental 45-Day Undersea Saturation Dive at 205 Feet. Sealab II Project Group* (D 210. 15:ACR-124) Washington, 1967. 428 pp., illus.

Presents the report on the Sealab II operation which was conducted between August 28 to October 14, 1965. 3,000 feet off Scripps Pier at La Jolla, California in a depth of water of 205 feet. Sealab II operations conducted by the Office of Naval Research as a part of the man-in-the-sea task of the Deep Submergence Systems Program was an interdisciplinary investigation into the usefulness of ocean floor habitation by the measurement of the ability of man to do useful work while living as a saturated diver in equilibrium with the ocean-floor pressure.

1000a *Project SeaLab Summary Report. An Experimental Eleven-Day Undersea Saturation Dive at 193 Feet. Sealab I Project Group.* (ONR Report ACR-108). Washington, Office of Naval Research, 1965. 58 pp., illus., diagrams.

Promenades d'Archeologie Sous-Marine. See entry 309.

1001 PROPOSTO, C. del. *Bateaux Sous-Marins a Grande Vitesse Sous l'Eau.* With a project of M. C. Laurenti. Brussells, Mertens, 1910. 107 pp., diagrams.

Proyecto de Codificación del Empleo del Submarino en la Guerra. See entry 205.

1002 PUGH, Marshall. *Commander Crabb.* London, Macmillan, 1956. 165 pp., illus. LC 56-59118.

1003 ________. *Frogman: Commander Crabb's Story.* NY, Scribners, 1956. 208 pp., illus. LC 56-11868.

Mysterious disappearance of CDR Lionel P. K. Crabb while on an underwater demolition mission.

Q

Q Boat Adventures. . . . See entry 37.

Q Ships. See entries 182-185.

Q-Ships and Their Story. See entries 218, 219.

Quarante-Deux Semaines de Guerre Sous-Marine. See entry 1077.

Quatre Annees de Guerre Sous-Marine. See entry 1408.

Queen Mary (Ship). See entry 406.

1004 QUILICI, Folco. *The Blue Continent.* Photos by Quilici and Girogio Ravelli. NY, Rinehart, 1954. 246 pp., illus. LC 54-8252.

Red Sea diving expedition to investigate the marine biology. 1952 and 1953.

1005 ________. *Mala Kebir: L'Avventura dei Subacquei di Sesto Continente alla Conquista di un Primato Mondiale.* Photos by Girogio Ravelli and the author. Rome, Casini, 1955. 153 pp. illus. LC 56-46754.

De Quinta Essentia. See entry 338.

R

1006 RABEAU, A. *Les Submersibles,* with Adolphe Laurens. Preface by ADM. Lacaze. Paris, Colin, 1925. 214 pp., illus., diagrams. bibliography. LC 26-7548.

1007 RACKL, Hanns Wolf. *Diving Into the Past: Archaeology Under Water.* Translated by Ronald J. Floyd. NY, Scribner, 1970. 292 pp. illus., maps.

There was also a 1968 edition.

Radar. See entry 85.

1008 RADIGUER, Charles Felix Adolphe, 1875- . *La Navigation Sous-Marine.* Paris, Doin, 1911. 361 pp., illus., diagrams, bibliog. LC 11-17664.

1009 RADLOFF, Roland and Robert HELMREICH. *Groups Under Stress: Psychological Research in Sealab II.* NY, Appleton, 1968. 259 pp., illus. ISBN 0-390-72600-1. LC 68-19962.

Rätsel der U-Bootswirkung. See entry 614.

Raiders of the Deep. See entries 1232, 1233, 1234.

1010 RALPHSON, G. Harvey. *Boy Scouts in the North Sea: or, Mystery of U-13.* Chicago, Donahue, 1915. 252 pp., frontis.

Juvenile fiction.

1011 RAMLOW, Gerhard, 1904- . *Ausgelaufen Westwärts: U-Boots-Schicksale von Rüben und Drüben.* Potsdam, Voggenreiter, 1937. 143 pp., illus., bibliog. LC 51-53931.

Ramskill, E. A. See entry 1392.

Randle Ford, W. T. See entry 689.

Rapier, HMS (fictional). See entry 293.

Rapport, S. B. See entry 1483.

Rapports de Mer Sur les Pertes de Guerre Subies par la Marine de Commerce Norvégienne. See entry 926.

Rasche, Herbert. See entry 687.

1012 RASCOVICH, Mark. *The Bedford Incident.* NY, Pocket Books, 1964. (c. 1963) 291 pp., illus.

Originally published by Atheneum in 1963.

Novel re USS Bedford versus a Soviet submarine. Cold War suspense story.

1013 RATHBONE, Alfred Day, 1897- . *He's In the Sub-Busters Now.* NY, McBride, 1943. 224 pp., frontis, illus. LC 44-465.

Ravelli, Giorgio. See entries 1004, 1005.

Raw, Sir Sydney. See entry 25.

1014 RAY, Carleton. *The Underwater Guide to Marine Life* with Elgin Ciampi. NY, Barnes, 1956. xiii, 338 pp., illus., maps, profiles, bibliog. LC 56-5558.

1015 RAYNER, D. A. *The Enemy Below.* NY, Holt, 1957. 191 pp. LC 56-11164.

Fiction. WW II Battle of the Atlantic. US destroyer versus German U-121. Was made into a successful movie.

Razvitie Minnogo Oruzhiia v Russkom Flote. See entry 1082.

Ready for Sea. See entry 938.

The Real Book About Submarines. See entry 379.

The Realm of the Submarine. See entry 233.

The Rebel Shore. See entry 854.

1016 REBIKOFF, Dmitri 1921- . *L'Aviation Sous-Marine.* Paris, Flammarion, 1962.

1017 ________. *Exploration Sous-Marine.* Paris, Arthaud, 1952. 211 pp., illus. LC 53-25670.

1018 ________. *Free Diving.* Preface by Yves Le Prieur. American ed. edited by Albert Vander-Kogel. NY, Dutton, 1956. 224 pp., illus. LC 56-6318.

Rechnitzer, Andreas. See entry 1146.

Receuil de Diverses Pieces Touchant Quelques Nouvelles Machines. See entry 949.

The Red Fleet in the Second World War. See entry 665.

The Red Navy. See entry 495.

Red Sea. See entries 1004, 1005, 1056, 1057.

Red Star Rising at Sea. See entry 498.

Redoutable (French submarine). See entry 764.

Reed, Edward James, 1830-1906. See entries 462, 463.

The Reefs of Taprobane. See entry 227.

1019 REEMAN, Douglas. *Dive in the Sun.* NY, Putnam, 1961. 257 pp. LC 61-14780.

Novel re midget submarine mission in WW II to blow up a floating dock on Adriatic Coast prior to Allied invasion of Italy.

1020 ________. *With Blood and Iron.* NY, Putnam, 1965. 258 pp. LC 65-10858.

Novel. WW II submarine warfare in the Atlantic. Fictional German submarine U-991.

1021 REES, Ed. *The Seas and the Subs.* NY, Duell, 1961. Toronto, Longmans, 1961. 253 pp., plates. LC 61-11963.

Regierung Doenitz. See entry 802.

1022 REGNAULT, J. *Le Sous-Marin, sa Situation dan le Droit des Gens.* n.p., Domat-Montchrestien, 1934.
Consideration of submarine warfare in international law.

1023 REHDER, Jacob. *U-Bootsfallen.* Munich, Lehmann, 1935. 159 pp., frontis, illus., plates. LC 36-6450.

Reichsleitung und U-Boot-Seinsatz 1914-bis 1918. See entry 68.

1024 REINHARD, Wilhelm. *Wir an den Maschinen: Kriegsfahrten auf und Unter See.* Leipzig, Hase & Koehler, 1940. 256 pp., illus., plates, ports, diagrams. LC 40-22830.

Rendezvous. See entry 640.

Rendezvous, and Other Long and Short Stories. See entry 641.

Rendezvous by Submarine. See entry 661.

1025 *Report and Hearings of Special Subcommittee on Proposed Undersea Warfare Laboratory, Los Alamitos, California.* 90th Congress, 1st Session, July 12-22, 1967. 4023-4185 pp., illus. (Armed Services Committee. House. Washington, D.C. Papers 90th Congress. Y4.Ar 5/2a: 967-68 #17.) LC 67-62184.

Report . . . Deep-Water Diving. See entry 516.

Report of the Commissioner Concerning Purchase of Submarines. See entry 187.

Report on Salvage Operations; Submarine S-51. See entry 374.

Report Upon Experiments and Investigations to Develop a System of Submarine Mines for Defending the Harbors of the United States. See entry 6.

1026 *Report Upon Trials With Submarine Mines Executed Jointly by Sweden, Norway and Denmark. 1874-1876.* Translated from the Danish by C. W. E. Oxholm. (U.S. Engineer School "Professional Papers" vol II no. viii) Willets Point, Printed on the Battalion Press. 1st Class Privates Beck and Phillips, printers, 1883. 109 pp., plates. LC 13-11680.

1027 *Report Upon Trials with Submarine Mines Executed Jointly by Sweden, Norway and Denmark. 1874-1876.* Translated from the Danish by C. W. E. Oxholm. (Engineer School. Professional Papers. vol. II, no. VIII) Willets Point, NYH, Printed on the Battalion Press, 1st Class Privates Beck and Phillips, printers, 1883. 109 pp., plates. LC 13-11686.

Rescue. See entries 1200, 1266.

The Rescuer. See entry 807.

Research submarines. See entries 284, 663, 780, 830.

Research Submersibles in Oceanography. See entry 829.

Return from the Deep. See entry 1266.

Revolutionary War. See entry 617.

1028 LA REVUE MARITIME. *Les Sous-Marins.* Paris, La Revue Maritime, 1955.

The June 1955 issue was a special issue concerning submarines.

Rezka Metallov Pod Vodoi i Raschistka Rusel. See entry 708.

1029 RIBERA JORDA, Antonio, 1920- . *La Conquista de las Profundidades.* with the collaboration of Mario Lleget Colomer. Barcelona, Editorial Hispano-Europea, 1955. 186 pp., illus., ports. LC 56-26105.

1030 ________. *La Exploración Submarina.* Barcelona, Editorial Seix Barral, 1956. 108 pp., illus., bibliog. LC A 56-5641.

1031 ________. *Guia Submarina de la Costa Brava.* Preface by Jose Pla. Barcelona, Ediciones Destino, 1956. 167 pp., illus., bibliog. LC 57-34733.

1032 ________. *Los Hombres-Peces: Exploración y Arqueologia Submarinas.* Preface by Luis Pericot. Map by Jacques Yves Cousteau. Photographs by Eduarado Admetilla and Clemente Vidal. Barcelona, Editorial Juventud, 1956. 206 pp., illus. LC 57-27804.

1033 ________. *Els Homes-Peixos: Exploració i Arqueologia Submarines.* Prefaces by Luis Pericot and Salvador Espiru. Barcelona, Editorial Selecta, 1954. 246 pp., illus. LC 55-22179 rev.

1034 RICHTER, Heinrich. *Otto Weddigen: Ein Lebensbild.* Bielefeld & Leipzig, Velhagen & Klasing, 1915. 149 pp., frontis., illus., facsim. LC 21-12406.

1035 RICKOVER, Hyman G. *Eminent Americans. Namesakes of the Polaris Submarine Fleet.* (92nd Congress, 2nd Session. House Doc. No. 92-345.) Washington, 1972. 316 pp., illus. appendix, index. (Superintendent of Documents Stock Number 5271-00315.)

41 biographical sketches of Polaris missile submarine namesakes. Photos of submarines. Appendix is a data table: "Polaris Fleet Ballistic Missile Submarine Shipbuilding Program."

Rickover, Hyman G. See also entries 66, 117, 118, 287, 606, 930 and 1281.

1036 RIESEBERG, Harry Earl. *I Dive for Treasure.* NY, McBride, 1942. 331 pp., frontis., plates, maps. LC 42-14791.

1037 ———. *Diving for Sunken Treasure.* Editorial assistance by Robert Wells. Illustrated by Jack Coggins. 64 pp., illus. LC 55-3563.

Rig for Depth Charges! See entry 590.

Riou, Edouard. See entry 1410.

1038 RITTER, Paul. *Ubootsgeist: Abenteuer und Fahrten im Mittelmeer,* nach kriegstagebuchern von maschinistenmaat Paul Ritter, herausgegeben von Gerd Block. Leipzig, Koehler, 1935. 246 pp., illus., maps, plates, ports. LC 36-6113.

Ritter der Sieben Meere. See entry 13.

Ritter der Tiefe. See entry 1235.

Ritter der Tiefe—Graue Woelfe. See entry 605.

Rivoire, Jean, 1929- . See entries 547, 751, 752.

Robert Fulton and the Submarine. See entry 951.

1039 ROBERTS, Fred M. *Basic Scuba: Self-Contained Underwater Breathing Apparatus.* Foreword by Edgar End. Edited by Eugene V. Connett. Second edition. NY, Van Nostrand, 1963. xxii, 488 pp., illus. LC 64-1105.

1040 ROBERTSON, Morgan. *Down to the Sea.* NY and London, Harper, 1905. 312 pp.

Fiction with Russo-Japanese War and "new-fangled" submarines as part of this science fiction tale. The submarines described are pseudo-Holland type A-boats. Pretty fair reading.

1041 ROBERTSON, Terence, 1921- . *Golden Horseshoe.* Preface by Sir George Creasy. London, Evans, 1955. Toronto, British Book Service, 1955. 210 pp., plates. LC 56-172.

Same as *Night Raider of the Atlantic.*

1042 ________. *The Golden Horseshoe.* London, Evans, 1966. Toronto, Copp, 1966. 159 pp., plates.

1043 ________. *Jagd auf de Woelfe.* (Hunting the wolves). Hamburg, Stalling, 1960. 224 pp., illus.

British ASW activity in Atlantic. The English edition bore the title *Walker, R.N.*

1044 ________. *Le Loup de L'Atlantique.* Paris?, Amiot-Dumont, 1955.

1045 ________. *Night Raider of the Atlantic: Saga of the German Submarine the Golden Horseshoe and her Daring Commander, Otto Kretschmer.* Preface by Sir George Creasy. NY, Dutton, 1956. NY, Ace, 1955. 256 pp. LC 56-5013.

The Golden Horseshoe was a symbolic device which Kretschmer had painted on the conning tower of his submarine U-99.

1046 ________. *Night Raider of the Atlantic.* NY, Ballantine, 1974. 192 pp., illus.

A reissue of their 1955 edition which was published under the title of *The Golden Horseshoe.*

1047 ________. *The Ship With Two Captains.* Forewords by Jerauld Wright and N. L. A. Jewell. 256 pp., illus., plates. LC 57-5345.

Same. London, Evans, 1957. Toronto, British Book Service, 1957. 192 pp., plates.

1048 ________. *Der Wolf im Atlantik.* Munich, Welsermuehl, 1955? 370 pp., illus.

Kretschmer and the U-99. German version of *The Golden Horseshoe.*

Robinson, E. H. See entry 1120.

1049 ROBINSON, William Morrison, Jr. *The Confederate Privateers.* New Haven, Yale University Press, 1928. xvi, 372 pp., illus. LC 28-28399.

An important work for study of Confederate submarines.

1050 ROCHOLL, Erich, 1885- . *Die Frage der Minen im Seekrieg vor Ihrer Beandlung auf der II. Haager Konferenz.* Borna-Leipzig, Noske, 1910. 47 pp. LC 44-43245.

Submarine mining as considered at the 2nd International Peace Conference at the Hague in 1907.

1051 ________. *Die Frage der Minen im Seekrieg auf der Grundlage des Haager Abkommens. . . .* Marburg, Ebel, 1910. 154 pp. LC 38-29314.

International law regarding submarine mining.

1052 ROCKWOOD, Roy, *pseud. Under the Ocean to the South Pole: or, The Strange Cruise of the Submarine Wonder.* NY, Cupples and Leon, 1907. 248 pp., frontis., plates. LC 7-24774.

Juvenile fiction in the "Great Marvels" series.

1053 ROEBLING, John A. Sons co. *Wire-Roping the German Submarine: The Barrage that Stopped the U-Boat.* Trenton, N.J., Roebling's, 1920. 37 pp., frontis., illus., map. LC 20-9034.

Extracts from an article "The Barrage that Stopped the U-Boat" by Reginald R. Belknap. *Scientific American* March 15-22, 1919. p. 7-9.

1053a ROESSLER, Eberhard. *Geschichte des Deutschen Ubootbaus.* (History of German Submarine Construction). Munich, Lehmann, 1974. 448 pp., 88 pictures, 200 tables, 200 plans and sectional drawings.

Covers the submarines of WWI, WWII, and the present time.

Author is an engineer who has had a long involvement with submarine construction. This is the definitive book on the topic.

1054 ________. *U-Boottyp XXI.* 2nd rev. ed. Munich, Lehmann, 1968. 160 pp., 34 illus., 64 drawings and schematics.

A technical treatise on the Walter Type XXI high-capacity battery, attack U-boat. This was the type which was ready at the end of World War II, but which never got into action. Excellent source book.

1055 ________. *U-Boottyp XXIII.* Munich, Lehmann, 1968.

116 pp., 32 illus., 38 drawings and schematics.

A treatise on the Walter Type XXIII high-capacity battery, coastal U-boat; with complete technical facts.

Roessler, Eberhard. See also entries 735, 922.

Roger Bacon's Letter Concerning the Marvelous Power of Art and of Nature. . . . See entry 42a.

1056 ROGHI, Gianni, 1927- . *Dahlak: With the Italian National Underwater Expedition in the Red Sea,* with Francesco Baschieri. Translated by Priscilla Hastings. Edited by Eleanor Brochett. Fair Lawn, N.J., Essential Books, 1957. 280 pp., illus. LC 57-13644.

1057 ________. *Dahlak: Con la Spedizione Nazionale Subacquea in Mar Rosso,* with Francesco Baschieri. 1st ed. Milan, Garzanti, 1954. 281 pp., illus. LC 54-26992.

1058 ROHWER, Jürgen. *Das Deutsche Bild der Russischen und Sowjetischen Marine.* (The German picture of the Russian and Soviet Navies). Frankfurt, 1962.

1059 ________. *66 Tage unter Wasser. Atom U-Schiffe und Raketen.* Oldenburg and Hamburg, Stalling, 1962. 88 pp., illus.

1060 ________. *Die Sowjetische U-Bootswaffe in der Ostsee, 1939-1945.* (Soviet submarine warfare in the Baltic, 1939-1945). Frankfurt, 1956.

1061 ________. *U-Boote: Eine Chronik im Bildern.* Hamburg, Stalling, 1962. 96 pp., illus.

A pictorial chronicle of German U-boats from 1935-1945.

1062 ________. *Die U-Bootserfolge der Achsenmächte, 1939-1945.* Munich, Lehmann, 1968. 376 pp., tables.

"The Submarine Successes of the Axis 1939-1945." Detailed information including Axis U-boat losses with the locations, dates and times of sinkings. Tables and listings of all U-boat successes with date, time and particulars. Lists all sunk, participating and surviving ships, and their commanders. A monumental work, twenty years in the making.

1063 ROWHER, Jürgen and G. HUMMELCHEN. *The War at Sea.* 1974. 650 pp., illus.

Roman Galley Beneath the Sea. See entry 387.

The Romance of Submarine Engineering. See entry 254.

The Romance of the Submarine. See entry 669.

1064 ROMANOVSKY, Vsevolod, 1912- . *L'Homme-Poisson et se Conquêtes.* Paris, Stock, Delamain et Boutelleau, 1955. 105 pp., illus. LC 55-57034.

Oceanographic research.

1065 ROMAT, Êtienne, 1914- . *Un Sous-Marin Attaque.* Preface by ADM. Barjot. Paris, Gigord, 1946. 63 pp., illus., maps. LC 49-15405 rev.

WW II submarine operations.

1066 ROMIG, Mary F. *Fatal Submarine Accidents: A Bibliography, 1900-1965.* Santa Monica, Calif., Rand Corporation, 1966. 22 pp.

Romulo, Carlos P. See entry 661.

1067 RONDELEUX, Marcel, 1870- . *L'Apogée de la Guerre Sous-Marine, Journal d'un Commandant D'Escadrille de Patrouille, 1917-1918.* Paris, Les Editions de France, 1937. 241 pp., map. LC 38-125.

1068 RØNNEBERG, Harald B. M. *Det Norske Ubåtvåpen: 1909-1959.* Bergen, Eides, 1959. 127 pp., illus., tables.

History of the Norwegian submarine force, 1909-1959.

1069 ROSCOE, Theodore. *Pigboats.* NY, Bantam Books, 1958. 449 pp.

Abridged paperback version of *U.S. Submarine Operations in WW II.*

1070 ________. *Le Service Silencieux.* Paris, France-Empire, 1955.

1071 ________. *United States Submarine Operations In World War Two.* Annapolis, U.S. Naval Institute, 1949. xx, 577 pp., illus., plates, maps. LC 50-5198.

Large format book illustrated by Fred Freeman. Author had access to submarine patrol reports and official information.

1072 ROSE, Hans. *Auftauchen! Kriegsfahrten von U-53.* With 20 pictures by Claus Bergen. Essen, Essener Verlagsanstalt, 1939. 315 pp., illus., maps. LC 50-49152.

WW I submarine operations.

1073 ROSKILL, Stephen Wentworth. *The Secret Capture*. London, Collins, 1959. Toronto, Collins, 1959. 161 pp., plates, maps.

Story of the capture of U-110 by British forces.

1074 ________. *Das Geheimnis um U-110*. Frankfurt am Main, Bernhard & Graufe, 1960. 147 pp., illus.

German edition of *The Secret Capture*.

Rosner, Charles. See entry 46.

Ross, Alan. See entries 310, 312.

1075 ROSS, Frank, Jr., 1914- . *Undersea Vehicles and Habitats: The Peaceful Uses of the Ocean*. NY, Crowell, 1970. 183 pp., illus. ISBN 0-690-84416-6. LC 76-106577.

For grades 5 through 8.

Ross, Vesey. See entries 574, 575.

1076 ROST, D. and R. HERVET. *Un Sous-Marinier*. Paris, France-Empire, 1960.

Roth, Herb. See entry 160.

Rough Notes . . . on Electricity as Applied to Submarine Mining. . . . See entry 212.

Rough Notes . . . on Submarine Mining . . . See entry 211.

1077 ROUSSEAU, Anselme. *Quadrante-Deux Semaines de Guerre Sous-Marine*. Paris, Alcan, 1918. 22 pp., 4 diagrams. LC-20-11669.

1078 ROUTLEDGE, Robert. *Discoveries and Inventions of the Nineteenth Century*. London & NY, Routledge, 1896. 717 pp., illus.

Royal Mail Lines. See entry 171.

Royal Oak, HMS. See entries 819, 998, 999.

Rubis (French submarine). See entry 996.

Rudolph, Norman Guthrie. See entry 654.

1079 RUDUITSKLY, M. A. "Underwater Ships of Science, USSR" from *Tekhuik Molodezhi* No. 7, July 1966. Translation published April 5, 1967. 21+7 pp., illus. Washington (?) Joint Publications Research Service. Series 8267-40527. (TT: 67-31171).

1080 RUGE, Friedrich. *Torpedo-und Minenkrieg.* Munich & Berlin, Lehmann, 1940. 67 pp., illus., map, diagrams. LC 43-15262.

Rukovodstvo Dlia Vodolaza. See entry 736.

Run Silent, Run Deep. See entry 73.

1081 RUSHMORE, David B. "Bibliography of the Literature of Submarines, Mines and Torpedoes, 1873-1917." Schenectady, *The General Electric Review,* 1917.

1082 RUSSIA. TSentralnyi Gosudarstvennyi Arkhiv Voenno-Morskogo Flota. *Razvitie Minnogo Oruzhiia v Russkom Flote.* Edited by A. A. Samorov. Moscow, 1951. xxvii, 349 pp., illus., maps. LC 52-40219.

1083 RUSSIA. Armiia. Glavnoe Voenno-inzhenernoe Upravlenie. *Nastavienie po Vodolaznemu Delu.* Moscow, 1937. 119 pp., illus. LC 50-48688.

Russian army diving manual.

Russian Arctic. See entry 549.

Russia at War: 1941-1945. See entry 1445.

The Russian Fleet in the First World War. See entry 526.

Russian Navy. See entries 503, 596, 665, 675, 676, 709, 719, 721, 723, 724, 725, 811, 868, 869, 872, 874, 906, 953, 968, 974, 990, 994, 1058, 1060, 1447, 1480.

The Russian Navy in the Russo-Japanese War. See entry 709.

The Russian Navy in War and Revolution. . . . See entry 503.

Russian Research Submarines. See entry 1079.

Russian Sea Power. See entry 386.

Russian Submarines. See entries 113, 121, 145, 174, 367, 386, 494, 635, 658, 720, 946, 1012, 1060, 1086, 1088, 1128, 1253, 1269, 1322, 1492.

Russian titles have been transliterated into English from the Cyrillic alphabet.

The Russians at Sea. See entry 1480.

The Russo-German War: 1941-45. See entry 1112.

Russo-Japanese War, 1904-05. See entries 674, 709, 994, 1040, 1252, 1447.

S

S-4 USS (SS 109). See entries 356, 370, 1308.

S-51 USS (SS 162). See entries 356, 370, 371, 372, 374, 375.

S-51: Par le Fond. See entry 375.

S-54: Stories of the Sea. See entry 376.

Safe Submarine Vessels. . . . See entry 739.

Safety and Operational Guidelines for Undersea Vehicles. See entry 830.

Sailfish, USS (SS 192). See entry 1266.

Sailors, Subs, and Señoritas. See entry 867.

1084 ST. LAWRENCE, William P. *Submarines.* (Science Service. Science Program Series) Garden City, Doubleday, 1959. 64 pp., illus., chart.

Saissy, A. See entry 414.

Salvage. See entries 53, 138, 172, 209, 254, 283, 288-292, 324, 356, 370, 371, 372, 374, 375, 497, 537, 803, 807, 843, 844, 851, 852, 1121, 1122, 1237, 1394, 1456, 1457, 1467.

Samarov, A. A. See entry 1082.

1085 SAND, George X., 1915- . *Skin and Scuba Diving.* (A Popular Mechanics Press Book) NY, Hawthorn Books, 1964. 123 pp., illus. LC 64-19489.

Sandburg, N. Elaine. See entries 1341, 1342.

1086 SAPUNAR, Zorislav. *Podmornica.* Belgrade, Tehnicka

Knjiga, 1952. 185 pp., illus. LC 54-43790.

Saucer in the Sea. See entry 1116.

1087 SAUER, Herbert. *Die Höllenmaschine im U-Boot.* Berlin, Scherl, 1928. 145 pp., frontis., port., plates. LC 29-4062.

1088 SAUNDERS, Malcolm George, ed. *The Soviet Navy.* NY, Praeger, 1958. London, Weidenfeld, 1958. Toronto, Ambassador, 1958. 340 pp., illus., plates, maps. LC 58-12091.

There are about five pages of this book devoted to subject of Soviet submarines.

1089 SAUSSAYE et PEAN, L. de la. *La Vie et les Ouvrages de Denis Papin.* Paris, Blois, 1869.

Savill, Mervyn. See entry 856.

Sazo, Serge de. See entry 411.

Du Scaphandre et de son Emploi a Bord des Navires. See entry 354a.

1090 SCHAEFFER, Heinz. *Geheimnis um U-977.* Buenos Aires, Editorial Prometheus, 1950. 299 pp., illus.

1091 ________. *U-Boat 977.* Foreword by Nicholas Monsarrat. NY, Norton, 1952. 260 pp., map endpapers. LC 53-6114.

Same. Translated by George Eclinger and Godfrey Wilson. Preface by the Earl of Cork and Orrery. Foreword by Nicholas Monsarrat. London, Kimber, 1952. 207 pp., plates. LC 65127.

1091a ________. *U-977. 66 Tage Unter Wasser.* Weisbaden, Limes, 1950. 264 pp., illus.

Schaeffer was captain of U-977 and rather than surrender when Hitler capitulated he took the submarine to South America. This is the story of the voyage.

1092 SCHAEFFER, Karl Ernst, 1912- . *Environmental Effects on Consciousness.* (First International Symposium on Submarine and Space Medicine. 1958. New London). NY, Macmillan, 1962. 146 pp., illus. LC 61-9079.

1093 SCHAFFERT, Arthur. *The Aquanauts.* NY, Coward, McCann, 1971.

Juvenile.

1094 SCHARFF, Robert. *The Collier Quick and Easy Guide to*

Skin and Scuba Diving. NY, Collier Books, 1963. 128 pp., illus. LC 63-17395.

1095 SCHARFENBERG, Horst. *Nautilus 90 Grad Nord: Atom-U-Boote Erobern die Meere.* Stuttgart, Thienemanns, 1959. 190 pp., illus.

1096 SCHEER, Reinhardt, 1863-1928. *Deutschlands Hochseeflotte im Weltkrieg.* Berlin, Scherl, 1920. 524 pp., port., illus., plans. LC 33-5890.

1097 ————. *Germany's High Sea Fleet in the World War.* London, Cassell, 1920. 375 pp., port., maps, plans. LC 20-8001.

1098 SCHEFFEL, Fritz, 1889- . *Der Brandtaucher: Wilhelm Bauer Erfindet das U-Boot: Die Geschichte eines Lebenskampfes.* Leipzig, Koehler & Amelang, 1936. 271 pp., plates, ports. LC 37-14249.

1099 SCHEIDNAGEL y SERRA, Leopoldo. *Extracts from a Treatise Upon Defensive Submarine Mining. . . .* Madrid, Corps of Engineers, Spanish Army, 1880. Translated by LT. Fred V. Abbot, Corps of Engineers. U.S.A. (U.S. Engineer School. Professional Papers. Vol. I No. II) Willets Point, N.Y.H., Printed on the Battalion Press, Sergt. Carmichael & Pvt. Beck, Printers, 1881. 85 pp., 3 plates. LC 13-11689.

1100 SCHELIHA, Viktor Ernst Karl Rudolf von, 1826-1899. *A Treatise on Coast-Defense Based on the Experience Gained by Officers of the Corps of Engineers of the Army of the Confederate States, and Compiled from Official Reports of Officers of the Navy of the United States.* London, Spon, 1868. xviii, 326 pp., frontis., illus., plus 15 folding plates which include maps and plans. LC 18-23651.

1101 SCHENCK, Hilbert Van Nydeck, 1926- . *Shallow Water Diving and Spearfishing,* with Henry Kendall. Illustrated by Vincent Vitale. Cambridge, Md., Cornell Maritime Press, 1954. 251 pp., illus. LC 53-9687.

1102 ————. *Skin Diver's and Spearfisherman's Guide to American Waters. Where to Go and What you Will Find When You Get There.* Illustrated by John E. Johnson. Cambridge, Md., Cornell Maritime Press, 1955. 58 pp., illus., bibliog. LC 55-10161.

1103 SCHEPKE, Joachim, 1912- . *U-Boot-Fahrer von Heute.*

Berlin, Deutscher Verlag, 1940. 131 pp., frontis., port., col. plates. LC 45-23056.

1104 SCHEURER, Adolf, 1883- . *Bewaffnete Handelsschiffe im Weltkriege.* Berlin, Springer, 1919. 78 pp. LC 50-41128.

Die Schiffe der Deutschen Kriegsmarine und Luftwaffe, 1939-1945. See entry 534.

Schlachtkreuzer im Nebel. See entry 440.

1105 SCHLEGEL, Karl. *Stander Z Vor! Erlebnisse als Torpedofünker vom Skaggerak bis Scapa Flow.* Textzeichnungen von Werner Chomton. Stuttgart, Union Deutsche Verlagsgesellschaft, 194? 66 pp., illus. LC A51-172 rev.

1106 SCHMALENBACH, Paul. *Die Geschichte der Deutschen Schiffsartillerie.* Herford, Germany. Koehlers Verlaggesell schaft, 1968. 202 pp., illus.

Schmidhammer, Arpad. See entry 1178.

Schomaekers, Günter. See entry 791.

Der Schrecken der Meere. See entry 1402.

1107 SCHULTZ, Paul. *Im U-Boot durch die Weltmeere.* Bielefeld und Leipzig, Velhagen & Klasing, 1931. 192 pp., plates. LC 32-1662.

1108 SCHULTZE-BAHLKE, Georg. *U-Boote.* Berlin, Schmidt, 1918. 203 pp., illus.

1109 SCHULZE, Franz L. C., 1856- . *Unsere Unterseeboote: Geschichtleiche Entwicklung und Wirkung dieser Modernsten Waffe.* Kassel, Brunnemann, 1915. 64 pp., plates, illus., map.

1110 SCHÜTTE, Otto F., comp. *Deutschlands U-Boote, der Schrecken Englands: Ein Bildwerk von der Deutschen U-Boot-Waffe.* Preface by Kurt Kiel. Minden, Koehler, 1940. 80 pp., chiefly illustrations. LC 49-57577.

Schwartz, Ira. See entries 1341, 1342.

Scientists Against Time. See entry 69.

1110a SCORPION (SSN 589). *United States Ship Scorpion. In Memorium.* Washington, U.S. Navy Department, 1969. vii, 64 pp., illus., partly colored. (D 201.2:Sco7). LC 78-602279.

1111 SCOTT, Ian. *My Wår at Sea.* London, Jenkins, 1943. 190 pp., plates. LC A44-529.

Scribleriad. See entry 180.

Scuba Diver. See entry 849.

The Sea and its Story. See entry 1120.

Sea Devils. See entry 129.

Sea Diver. A Quest for History Under the Sea. See entry 782.

Sea Hazard (1939-45). See entry 630.

The Sea Our Shield. See entry 396.

Sea Power in the Machine Age. See entry 151.

The Sea Searchers. See entry 1118.

Sea Warfare, 1939-45. See entry 274.

The Sea Wolves. See entry 431, 432.

The Sea Wolves. See entry 634.

Seaborne Trade. See entry 394.

Seadragon: Northwest Under the Ice. See entry 1173.

Seadragon, USS (SSN 584). See entry 1173.

Seahorse, HMS (fictional). See entry 1477.

Seal, HMS. See entry 1429.

Sealab. See entries 1117, 1442.

Sealab I. See entries 1000a, 1393.

Sealab II. See entries 952, 1000, 1009.

The Seas and the Subs. See entry 1021.

1112 SEATON, Albert, 1921- *The Russo-German War: 1941-45.* NY, Praeger, 1971. 628 pp., maps. LC 70-130459.

Seawolf, USS (SS 197). See entries, 189, 429.

Sechs Jahre U-Bootfahrten. See entry 1160.

The Secret Capture. See entry 1073.

The Secret Invaders. See entry 1193.

Secret Mission Submarine. See entry 686.

The Secret Raiders. See entry 1481.

The Secret Service Submarine. See entry 1243.

Secretus Operibus Artis et Naturae, de. See entry 44.

Secrets de la Science Americaine. See entry 69.

Secrets of the Submarine. See entry 585.

Der Seekrieg in de Osteuropaeischen Gewassern. See entry 853.

Die Seeminen im Krieg. . . . See entry 97.

Segala, Renzo, 1905- . See entry 1404.

1113 SEIBT, Gustav, 1873- . *Die Aushungerung Englands, Eine Volkswirtschaftliche Untersuchung.* Berlin, Scherl, 1917. 48 pp. LC 24-10750.

1114 SELOW-SERMAN, K. E. *U-Boot-Abenteuer im Speergebeit.* Berlin, Scherl, 1917. 112 pp. LC 21-12691.

Sem-Jacobsen, Jens Helge, 1884- . See entry 1158.

Send Down a Dove. See entry 814.

Seraph, HMS. See entry 686.

Serpent of the Seas, The Submarine. See entries 251, 252.

Le Service Silencieux. See entry 1070.

Seven Miles Down. See entry 971.

The Shadow in the Sea. See entry 946.

Shallow Water Diving and Spearfishing. See entry 1101.

1115 SHANKLAND, Peter and Anthony HUNTER. *Dardanelles Patrol.* NY, Scribner, 1964. 192 pp., illus. LC 64-20053.

WW I period British submarine E-11.

1116 SHANNON, Terry. *Saucer in the Sea: The Story of the Cousteau Diving Saucer in Pacific Coast Waters.* Illustrated with photos, drawings and diagrams by Charles Payzant. San Francisco, Golden Gate, 1965. Toronto, Allen, 1965. 80 pp., illus. LC 65-11417.

Written for juvenile audience but valuable for its illustrations.

1117 SHANNON, Terry and Charles PAYZANT. *Project Sealab: The Story of the U.S. Navy's Man-in-the-Sea Program.* (Conducted by the Deep Submergence Systems Project of the Navy's

Special Projects Office in Conjunction with the Office of Naval Research.) San Francisco, Golden Gate, 1966. Toronto, Allen, 1966. 96 pp., illus. LC 66-7455.

1118 ________. and ________. *The Sea Searchers: Men and Machines at the Bottom of the Sea.* San Francisco, Golden Gate, 1968. 95 pp., illus. LC 68-22392.

Juvenile.

1119 ________. and ________. *Windows in the Sea: New Vehicles That Scan the Ocean Depths.* Chicago, Children's Press, 1973. 78 pp., illus. ISBN 0-516-08876-9. LC 73-9764.

Sharks and Little Fish. See entry 941.

1120 SHAW, Frank H. and E. H. ROBINSON. *The Sea and its Story: From Viking Ship to Submarine.* London, Cassell, 1910. 201 photos., 12 color plates, 121 drawings. Index of ships. LC 11-1708.

Historical survey. Good for the knowledge of the period, but somewhat dated at this time.

Shaw, Le Vergne. See entry 256.

Sheehan, E. A. See entry 1282.

1121 SHELFORD, W. O. *Sub Sunk: The Story of Submarine Escape.* London, Harrap, 1960. Garden City, Doubleday, 1960. 248 pp., illus., appendices. LC 60-52302.

With a foreword by B. W. Taylor.

1122 ________. *Subsunk.* NY, Popular Library, 1960. 240 pp., diagrams.

Paperback reprint.

1123 SHELNICK, Thomas. *Above and Below* with Ray Cristina. Garden City, Doubleday, 1965. London, Hale, 1966. 192 pp., illus. LC 65-22574.

Autobiography of a diver.

1124 SHENTON, Edward H. *Diving for Science: The Story of the Deep Submsersibles.* NY, Norton, 1973. 267 pp., illus. LC 74-90990.

Profile diagrams of various deep diving submersible craft. An appendix lists statistics and specifications for about 75 submersibles in existence.

1125 ________. *Exploring the Ocean Depths: The Story of the Cousteau Diving Saucer in the Pacific.* NY, Norton, 1968. 205 pp., illus., bibliog. ISBN 0-393-06363-1. LC 67-12450.

Shenton is an oceanographer who at the time of writing this book was associated with Westinghouse Undersea Division.

1126 SHERIDAN, Martin, 1914- . *Overdue and Presumed Lost: The Story of the USS Bullhead.* Francestown, N.H., Marshall Jones, 1947. xiii, 143 pp., illus. LC 47-6777.

1127 ________. same. Reissued 1959. Plus a memorial edition bound in leather.

The USS Bullhead (SS 332) was the last US submarine to be lost during WW II. Reported missing in August 1945 near Lombok Strait.

1128 SHERR, S. A. *Korabli Morskikh Glubin.* Moscow, 1955. 302 pp., illus., ports. LC 56-27216.

Shevlachov, H. See entry 1494.

Shilling, Charles W. See entry 1277.

The Ship with Two Captains. See entry 1047.

Ships and Sailors of the Red Navy. See entry 113.

Ships May Proceed. See entry 390.

Ships, Shoals, and Amphoras. See entry 294.

Ships That Explore the Deep. See entry 1478.

Ships Under the Sea. See entry 365.

Shipwrecks and Archaeology. See entry 1250.

1129 SHIRREFFS, Gordon D. *Torpedoes Away!* Philadelphia, Westminster Press, 1967. 206 pp., bibliog. LC 67-18730.

Boy's adventure fiction. WW II the fictional USS Grayfin in the Java Sea.

Shock Design of Shipboard Equipment. See entry 1389.

1130 SHPAKOVICH. F. A. *Vodolaznoe Snariazhenie.* Moscow, 1943. 111 pp., illus. LC 50-43123.

Shunk, Francis R. See entry 270.

1131 SHUTE, Nevil, *pseud.* *On the Beach.* NY, Morrow, 1957. London, Heinemann, 1957. Melbourne, 1957. 320 pp. LC

57-9158.

Produced as a movie under the same name. Much of the movie was was filmed off New Zealand with vessels of the New Zealand Navy. ADM. Lockwood served as technical advisor.

Sibley, K. M. See entry 385.

Sie Jagten Nippons Flotte. See entry 791.

1132 SIEBE, Henry. *The Conquest of the Sea: A Book About Divers and Diving.* NY, Routledge, 1874? xi, 299 pp., frontis., illus., plates. LC 18-4206.

Signals and Signaling. See entries 1201, 1204.

The Silent Service. See entries 208, 209.

The Silent World. See entries 260, 261.

1133 SILVERBERG, Robert. *Sunken History: The Story of Underwater Archaeology.* NY, Chilton, 1963. Toronto, Ambassador, 1963. 177 pp., illus. LC 63-7507.

1134 ________. *Treasures Beneath the Sea.* Illustrated by Norman Kenyon. (Whitman Book) NY, Golden Press, 1964. 90 pp., illus.

Juvenile.

Silversides, USS (SS 236). See entry 1268.

Silversides. See entry 1268.

1135 SIMPSON, Colin. *The Luisitania.* Boston, Little, Brown, 1973.

Simpson, G. W. G. See entry 673.

Sims, Lydel. See entry 527.

1136 SIMS, William Sowden, 1858-1936. *The Victory at Sea.* In collaboration with Burton J. Hendrick. Garden City, Doubleday, Page, 1920. xiii, 410 pp., port, map. LC 20-18578.

Admiral Sims was commander of U.S. Naval Forces operating in European waters during WW I.

Sims, William Sowden, 1858-1936. See also entry 183.

Simsadus. See entry 763.

Sinetskiy, P. See entry 1322.

Sink 'Em All. See entry 792.

Sinuous Courses. . . . See entry 64.

Six Years of Submarine Cruising. See entry 1161.

1600 Years Under the Sea. See entry 388.

60 Jahre Deutsche U-Boote, 1906-1966. See entry 603.

66 Tage Unter Wasser. See entry 1959.

Sjøforklaringer Over Norske Skibes Krigsforlis. . . . See entry 927.

Skate, USS (SSN 578). See entry 179.

1137. SKERRETT, Robert G. *The Evolution of the Submarine and How Far the Lake Type Solves the Problem.* Fort Monroe, Va., 1904. 24 pp., plates. LC 5-23007.

Reprinted from *Journal United States Artillery.* September-October, 1904. vol. 22, no. 2.

1138 ________. *The Submersible.* NY?, 1910. 37 pp., illus., map, plates. LC 10-13861.

Reprinted from the *Journal of the American Society of Naval Engineers.* May, 1910. vol. 22, no. 2, pp. 465-502.

Skin and Scuba Diver. See entry 128.

Skin and Scuba Diving. See entries 504, 1085.

Skin and Scuba Diving. See entries 48, 54, 78, 98-101, 107, 150, 152, 161, 162, 170, 196, 204, 227, 245, 258-263, 275, 278, 279, 300, 308-313, 333, 334, 343, 368, 388, 411, 422, 425, 443, 499, 500, 541, 547, 557, 561, 581, 678, 710, 712, 749, 767, 768, 782, 803, 812, 849, 945, 966, 967, 976, 1004, 1005, 1014, 1017, 1018, 1029, 1030-1033, 1039, 1056, 1057, 1064, 1083, 1094, 1101, 1102, 1133, 1134, 1179, 1180, 1181, 1213, 1214, 1215, 1217, 1244, 1245, 1249, 1250, 1251, 1275, 1407, 1462.

Skin Diver's and Spearfisherman's Guide to American Waters. See entry 1102.

The Skin Diver's Travel Guide. See entry 1251.

Skin Diving. See entrries 108, 339.

Skin Diving and Exploring Underwater. See entry 1209.

1139 SKOWRONNEK, Fritz Richard Bernhard, 1858- . *U-Deutschlands Fahrt.* Berlin, Janke, 1915. 161 pp.

1140 SKRODZKI, Bernhard. *Die Unterseebootfrage auf der Washingtoner Abrüstungskonferenz 1921/22.* Berlin, Dümmler, 1925. 88 pp. LC 26-7558.

The consideration of submarine warfare at the Washington Conference on the Limitation of Armament.

Skyirsky, David. See entry 720.

1141 SLEEMAN, Charles William. *Torpedoes and Torpedo Warfare.* Portsmouth, England, Griffin & co., 1880. 309 pp., illus., diagrams, plates.

Portsmouth, Simpkin, Marshall, 1889. 354 pp., illus., diagrams, plates.

1142 SLOSKY, Bill and Art WALKER. *Guide to the Underwater.* NY, Sterling, 1966. London, Oak Tree Press, 1966. 192 pp., illus. LC 66-16195.

Juvenile.

Smith, Douglas Hastie. See entries 915-917.

1143 SMITH, Gaddis. *Britain's Clandestine Submarines, 1914-1915.* New Haven, Yale University Press, 1964. 155 pp. LC 64-20937.

Reprinted Hamden, Connecticut, Archon Books, 1975.

Scholarly study of Britain's purchase of submarines in Canada from Vickers, Ltd., and the American firms of Lake Torpedo Boat Co. and Electric Boat Company, WW I era. The book was published as No. 4 in the Wallace Notestein Essay series.

Smith, Phil. See entry 812.

Smith, S. H., Jr. See entry 1390.

1144 SMITH, William Strother, 1857- . *Naval Inventions.* Washington, U.S. Government Printing Office, 1917. 10 pp., illus. LC 17-29016.

Article by a USN Captain which first appeared in the *Saturday Evening Post.* Discusses submarines and torpedoes.

1145 SNELL, Roy J. *The Flying Sub.* Chicago, Reilly & Lee Co., 1925. 282 pp.

Juvenile fiction.

1146 SNYDER, Dick. *The Trieste: The Story of the United States Navy's First Inner Space Ship.* Text and photos by Dick Snyder. Foreword by Andreas Rechnitzer, former scientist-in-charge of the USN bathyscaphe program. San Carlos, Calif., Golden Gate Junior Books, 1964. 80 pp., illus. LC 64-13269.

This is the second edition which includes technical data on *Trieste and Trieste II*. Preface includes information re the sunken *Thresher* and *Trieste II* involvement in the search. Juvenile, but text is lucid and the book is valuable for its many photographs and diagrams.

So War der U-Bootskrieg. See entries 167, 169.

1147 SOKOL, A. E. *The Imperial and Royal Austro-Hungarian Navy.* Annapolis, Naval Institute Press, 1968. 172 pp., illus., appendices.

1148 SOLAR CORREA, Eduardo, 1891-1935. *La Cuestión Submarına ante el Derecho Internacional.* 2nd ed. Santiago de Chile, Imprenta Cervantes, 1917. 128 pp. LC 42-35467.

Originally presented as a thesis at the Universidad de Chile.

1149 SOLOMON, Robert S., ed. *The Little David. C. S. S. David.* Moncks Corner, S.C., Berkeley Co. Tricentennial Committee, 1970. 44 pp., illus., bibliog.

Concerns Confederate submarines.

Solonevich, George. See entry 726.

Solórzano, Enrique. See entry 269.

Sommergibili. See entries 859, 860.

I Sommergibili della Seconda Guerra Mondiale. See entry 45.

1150 *Sommergibili in Guerra.* (Ufficio Storico della Marina Militare). Rome?, Ministero della Difesa-Marina, 1956. 215 pp.

Foreword by ADM. Giuseppe Fioravanso. WW II Italian submarines in the Mediterranean.

Sommergibili in Guerra. See entry 407, 1150.

I Sommergibili Italiana. See entry 329.

I Sommergibili Italiani 1895-1962. See entry 987.

Sonar. See entry 1356.

1151 SOPOCKO, Eryk K. S. *Orzel's Patrol: The Story of the Polish Submarine.* London, Methuen, 1942. Toronto, Saunders, 1942. xi, 146 pp., plates, ports. LC 42-4075.

The *Orzel*, a former USN submarine in Polish service during WW II was sunk, by mistake, in the English Channel by British ASW craft.

Sørøverens Faerd. See entry 31.

Sottomarini, Sommergibili e Torpedini. See entry 142.

1152 SOULE, Gardner. *The Greatest Depth: Probing the Seas to 20,000 Feet and Below.* NY?, Macrae Smith, 1970. 194 pp., illus. ISBN 0-8255-8350-0. LC 70-87988.

Story of the *Trieste* written for a juvenile audience.

1153 ________. *The Ocean Adventure: Science Explores the Depths of the Sea.* NY, Appleton, 1966. 278 pp., illus. LC 66-14956.

Underwater exploration for young readers.

1154 ________, ed. *Under the Sea: A Treasury of Great Writing About the Ocean Depths.* NY, Meredith, 1969. 347 pp., illus. LC 68-9516.

1155 ________. *Undersea Frontiers: Exploring by Deep-Diving Submarines.* Chicago, Rand-McNally, 1968. 253 pp., illus. LC 68025128.

Sous la Banquise. See entry 774.

Le Sous-Marin. . . . See entry 105.

Le Sous-Marin, Roi de la Mer. See entries 1258, 1259.

Le Sous-Marin, sa Situation dans le Droit des Gens. See entry 1022.

Un Sous-Marin Attaque. See entry 1065.

Les Sous-Marine Allemands devant New-York. See entry 79.

Les Sous-Marine et la Prochaine Guerre Navale. See entry 920.

Un Sous-Marinier. See entry 1076.

Sous Mariniers. See entry 545.

Les Sous-Marins. See entries 229, 285, 650, 722, 1028.

Les Sous-Marins Allemands. . . . See entry 753.

Les Sous-Marins du Soleil Levant. See entry 580.

Sous-Marins et Croiseurs Français. See entry 1236.

Les Sous-Marins et L'Angleterre. See entry 413.

Les Sous-Marins et la Guerre Actuelle. See entry 120.

Sous-Marins et Submersibles. . . . See entry 754.

Sous-Marins, Torpilles et Mines. See entry 755.

Sovetsky Flot v Voine s Yaponiei. See entry 470.

Sovetsky Voenna-Morskoi Flot. See entry 721.

The Soviet Fleet in the War with Japan. See entry 470.

Soviet Naval Developments. See entry 811.

Soviet Naval Power. See entry 990.

Soviet Naval Strategy. See entry 596.

The Soviet Navy. See entry 1088.

Soviet Navy. See entry 498.

Soviet Navy. See also Russian Navy.

The Soviet Sea Challenge. See entry 367.

Soviet Seapower in the Caribbean. See entry 1226.

Die Sowjet-Flotte im Zweiten Weltkrieg. See entry 974.

Die Sowjetische U-Bootswaffe in der Ostsee, 1939-1945. See entry 1060.

Spanish Civil War, 1936-1939. See entry 932.

Spanish Submarine Service. See entry 828.

Spanish Submarines. See entries 382, 824, 879, 880, 959, 962, and 1412.

Spearfishing in Britain. See entry 812.

Speirs, Jack. See entry 128.

The Spider Web. See entry 977.

1156 SPIEGEL von und zu PECKELSHEIM, Edgar, freiherr, 1885- . *The Adventures of the U-202: An Actual Narrative.* NY, Century, 1917. vii, 202 pp. LC 17-6334.

1157 ________. *U. Boat 202, the War Diary of a German Submarine.* Translated by Barry Domvile. London, Melrose, 1919. x, 170 pp. LC 20-3008.

1158 ________. *U 202: Krigsdagbok, Begyndt den 12th April*

19—, Sluttet den 30the April 19—. Translated by J. Sem-Jacobsen. Kristiania, Aschebourg, 1916. 120 pp. LC 17-1640.

1159 ________. *U-Boot im Fegefeuer.* Berlin, Scherl, 1930? 210 pp., frontis., plates, ports. LC 33-532.

________. See also entries 185, 1235.

1160 SPIESS, Johannes, 1888- . *Sechs Jahre U-Bootfahrten.* Berlin, Hobbing, 1925. 212 pp., frontis., plates, ports. LC 25-9411.

1161 ________. *Six Years of Submarine Cruising.* Washington, Office of Naval Intelligence, 1926. 130 pp., mimeo.

This is a translation of entry 1160. The NAS/NRC *Submarine Bibliography* misspells the name as Speiss in their citation.

1162 ________. *U-Boot-Abenteuer: Sechs Jahre U-Boot-Fahrten.* Berlin, Kolk, 1932. 184 pp., plates, ports. LC 32-25469.

An abridged reissue.

1163 ________. *Wir Jagten Panzerkreuzer, Kriegsabenteuer eines U-Boot-Offiziers.* Berlin, Steiniger, 1940, 1938. 207 pp., illus. LC 49-31956.

No matter how you slice it it is still *Sechs Jahre U-Bootfahrten.*

________. See also 681, 684.

1164 SPILHAUS, Athelstan Frederick, 1911- . *The Ocean Laboratory.* Mankato, Minn., Creative Education Society in Cooperation with the American Museum of Natural History, 1967. 112 pp., diagrams, glossary. LC 66-24868.

Juvenile.

1165 SPINDLER, Arno. *Der Handelskrieg mit U-Booten.* Berlin, Mittler, 1932-1941. 4 vols., illus., maps, bibliography. LC 32-13253.

A very important work for the study of WW I submarine operations.

1166 ________. *La Guerra al Commercio con i Sommergibili.* Rome, Instituto Poligrafico dello stato, Libereria, 1934- 3 vols. LC 49-58741 rev.

Italian translation of entry 1165.

1167 ________. *La Guerre Sous-Marine.* Translated by René Jouan. Paris, Payot, 1933- v. LC 35-16214.

French translation of entry 1165.

Spisok Korablei Russkogo. . . . See entry 872.

The Splinter Fleet of the Otranto Barrage. See entry 865.

Squailfish, USS (nickname). See entry 1266.

Squalus, USS (SS 192). See entries 53, 370, 807, 1266.

Srawley, J. E. See entry 1391.

SSSR-Velikaia Morskaia Derzhava. See entry 724.

Stability and Motion Control of Ocean Vehicles. See entry 8.

1168 STADTLÄNDER, Gerd, 1914- . *Die Verwendung von Minen im Seekrieg. Eine Völkerrechtliche Studie.* Dusseldorf, Dissertations-Verlag, 1938. x, 85 pp. LC 41-36028.

1169 STAFFORD, Edward Peary, 1918- . *The Far and the Deep.* Foreword by Charles A. Lockwood. NY, Putnam, 1967. London, Barker, 1968. 348 pp., illus. LC 67-23135.

Stalowe Rekiny. . . . See entry 1496.

1170 STAMBLER, Irwin. *The Battle For Inner Space: Undersea Warfare and Weapons.* NY, St. Martin's Press, 1962. 259 pp., illus. LC 62-8321.

USN nuclear submarines.

Stand by to Surface. See entry 70.

1171 STANDARD MOTOR CONSTRUCTION CO. *80 Foot Elco Submarine Chaser: Instructions, Care and Operation of Machinery Plant.* Jersey City, Standard Motor Construction Co., 1917. 203 pp., illus., plates. LC 17-25798.

Stander Z Vor! See entry 1105.

1172 STARCK, Walter A., 2nd. and Paul BRUNDZA. *Art of Underwater Photography.* Philadelphia, Chilton, 1966. 160 pp., illus. LC 66-23325.

Status of Research in Underwater Physiology. See entry 893.

Stead, Christiana. See entries 969, 970.

Steadfast, H.M.S. See entry 552.

1173 STEELE, George Peabody, 1924- . *Seadragon: Northwest Under the Ice.* NY, Dutton, 1962. 255 pp., illus. LC 62-7802.

1174 ________. *Vengeance in the Depths: The Nuclear Submarine.* NY, Dutton, 1963. Toronto, Clarke-Irwin, 1963. 183 pp., illus. LC 63-8599.

Juvenile.

1175 ________ and Herbert Julius GIMPEL, 1915- . *Nuclear Submarine Skippers and What They Do.* NY, Franklin Watts, 1962. Toronto, Ambassador, 1962. 140 pp. LC 62-9814.

Juvenile.

1176 STEEN STEENSEN, R. *Vore Undervandsbâde Gennem 50 Dr.* København, 1968.

Steinbrinck, Otto. See entry 602.

1177 STEINMETZ, Joseph A., comp. *Submarine Problems and Torpedo Defense.* Philadelphia, 1915. 96 pp., illus. LC 16-25811.

Reprints of periodical articles, including one about John P. Holland.

1178 STEINWÄGER, Leonhard. *U-Boot, Englands Tod!* Munich, Lehmann, 1918. 48 pp., illus., diagrams. LC 21-10834.

Illustrations by Arpad Schmidhammer.

1179 STELZNER, Hermann, 1884-1942. *Tauchertechnik, Handbuch für Taucher über den Bau und die Anwendung von Tauchergeräten aller Art, Lehrbuch für Taucheranwarter.* Lubeck, Coleman, 1943 (c 1931). 403 pp., illus., port. bibliography. LC 52-57914.

1180 STÉNUIT, Robert. *The Deepest Days.* Translated by Morris Kemp. NY, Coward-McCann, 1966. London, Hodder, 1966. Toronto, Longmans, 1966. 222 pp., illus., plates, maps. LC 66-10428.

1181 ________. *Vacances Chez Neptune.* Paris, Arts et Voyages, 1965.

1182 STEPHENS, Edward Carl, 1924- . *Blow Negative.* Garden City, Doubleday, 1962. Toronto, 1962. 466 pp., LC 62-8930.

Paperback edition: NY, Dell, 1963.

1183 ________. *The Submariner.* NY, Doubleday, 1973. 215 pp. ISBN 0-385-08884-1. LC 73-79715.

1184 ________. *Submarines: The Story of Underwater Craft from the Diving Bell of 300 BC to Nuclear-Powered Ships.* Illustrated by Jack Coggins. Cover by Lowell Hess. NY. Golden Press, 1959. 56 pp., illus. partly colored. LC 60-3881.

De Ster van 1572. See entry 890.

1185 STERLING, Forest J. *Wake of the Wahoo.* Foreword by Charles A. Lockwood. Philadelphia, Chilton, 1960. Toronto, Ambassador, 1960. 210 pp., plates. LC 60-8176.

The career of the USS Wahoo and her skipper Dudley "Mush" Morton in the Pacific during WW II.

Stevens, R. H. See entries 144, 318.

1186 STEWART, Harris Bates, 1922- . *Deep Challenge.* NY, Van Nostrand, 1966. 202 pp., illus. LC 66-27523.

Based on his book *The Global Sea.*

The Stick and the Stars. See entry 700.

Der Stier von Scapa Flow. See entry 433.

Stillson, George David, 1882- . See entry 1334.

1187 STOCKTON, Frank R. *The Great Stone of Sardis.* NY, Harper & Bros., 1898. Illustrated by Peter Newell. LC 8-15663.

Fictional account of a submarine. Jules Verne up-date.

1188 STOK, M. C., ed. *Wegduiken: De Nederlandse Onderzeedienst, 1906-1966.* Den Hag, Holland, Zuid-Hollandsche Uitgevers Maatschappij, 1966. 633 pp., illus., diagrams.

Issued to commemorate the 60th anniversary of the Dutch submarine service.

1189 STOKER, Henry Hew Gordon Dacre, 1885- . *Straws in the Wind.* London, Jenkins, 1925. 315 pp., frontis., plates, ports, maps, facsims. LC 25-14386.

WW I submarine operations. Turkish prisons.

Stoker, H. G. See also entry 215.

The Stolen Submarine. See entry 529 and 696.

Storage batteries. See entry 1336.

Storage Batteries for Submarines. See entry 1290.

Stories of Famous Submarines. See entry 940.

The Story of Our Submarines. See entry 135.

The Story of Submarines. See entry 1443.

The Story of the Submarine. See entries 111, 112.

The Story of the Submarine from the Earliest Ages to the Present Day. See entry 405.

The Story of the U-505. See entry 888.

1190 STOTHERD, Richard Hugh, 1828-1895. *Notes on Torpedoes, Offensive and Defensive*. Washington, Government Printing Office, 1872. 318 pp., illus., diagrams. LC 13-5892.

1191 STRANG, Herbert (*pseud.* of George H. Ely and C. J. L'Estrange). *Lord of the Seas*. NY, Hodder, n.d. London, Oxford, 1933. (Handy Library edition). 238 pp.

Strategy. See entry 1454.

Straws in the Wind. See entry 1189.

De Strijd Onder Water. See entry 804.

1192 STROH, Henri Charles, 1887- . *Mines et Torpilles*. Paris, Colin, 1924. 183 pp., illus., bibliog. LC 24-23182.

Stroh, Henri Charles, 1887- . See also entry 755.

1193 STRUTTON, Bill and Michael PEARSON. *The Beachhead Spies*. NY, Ace Books, 1958. 191 pp.

Published in London by Hodder and Stoughton with the title *The Secret Invaders*.

WW II in the European Theatre. Underwater demolition teams.

1194 STUBBS, Charles. *The Position of Enemy Merchantmen*. (In Grotius Society. *Problems of the War*. . . . London, 1916-19. vol. I, 1916, p. 19-22) LC A 21-996 Revised.

1195 *The Sub Base Gazette*. Groton, USN Submarine Base, 1940?- . A weekly newspaper with news of interest to submarine forces afloat and ashore. Published during WW II.

Sub Sailor. See entry 822.

Sub Sunk. See entry 1121, 1122.

Submarine. See entry 642.

Submarine! See entry 74.

Submarine (NavPers 16160). See entries 1359, 1360, 1361.

Submarine: A One Act Drama. See entry 801.

Submarine Adventures of Little Jinks. See entry 687.

Submarine Alone. See entry 552.

Submarine and Anti-Submarine. See entry 911.

The Submarine and Kindred Problems. See entry 1325.

The Submarine and Sea Power. See entry 607.

1196 SUBMARINE ARMOUR COMPANY. *Charter &c. of the Submarine Armour Company. Incorporated by the Legislature of the State of New York, April 2, 1838*. NY, 1838. 21 pp. LC 18-4205.

Submarine at Bay. See entry 835.

The Submarine at War. See entry 800.

Submarine Base. New London. See entries 321, 1328.

1197 *Submarine Bibliography*. 1954- . Edited by W. A. H. Brase. Braunschweig, 1954- . Annual. LC 54-35909.

The Submarine Boat Argonaut. See entry 742.

Submarine Boats. See entry 632.

Submarine Boys series. See entry 355.

Submarine Commander. See entry 156.

1198 SUBMARINE COOK BOOK COMMITTEE. *Dolphin Dishes*. Norfolk, Teagle and Little, Inc., 1959. 284 pp., illus.

Reprint of *Dolphin Dishes, The Submarine Cook Book with Favorite Recipes of the Families of the Submarine Force, U.S.N.* New London?, Published for the Benefit of the Navy Relief Society, 1952. 284 pp., illus. LC 52-68097.

Submarine Division Five. See entry 271.

Submarine Engineering. See entry 1423.

Submarine Engineering of Today. See entry 324.

Submarine Escape Training Experience, Summary of 25 Year Period. See entry 1346.

Submarine Fighter of the American Revolution. See entry 1420.

Submarine Flotilla. See entry 553.

The Submarine Force Library and Museum. See entry 1468.

Submarine Habitability. See entry 892.

The Submarine in the United States Navy. See entries 898, 899.

Submarine in the U.S. Navy. See entry 1382.

Submarine in War. See entry 883.

The Submarine in War and Peace. See entry 741.

The Submarine, Its Importance as Part of a Fleet. See entry 66.

The Submarine Library. See entry 475.

Submarine Material Guide. See entry 1369.

Submarine Medicine. See entries 95, 300, 343, 616, 718, 861, 862, 863, 892, 893, 895, 896, 1009, 1092, 1210-1212, 1337-1353.
This term includes physiology and psychology.

Submarine Medicine Practice. See entry 1337.

Submarine: Men and Ships of the U.S. Submarine Fleet. See entry 235.

Submarine Navigation Past and Present. See entry 163.

The Submarine Peril. See entries 683, 684.

Submarine Pioneer. See entry 886.

Submarine Problems and Torpedo Defense. See entry 1177.

1199 *Submarine Recognition Manual.* (NavPers 10011) Washington, U.S. Government Printing Office, 1965 (?) 38 pp., illus., diagrams.
Supersedes the earlier manual—OPNAV P31-102.

Submarine Recognition Manual. See entry 1362.

Submarine Rendezvous. See entry 654.

Submarine Rescue. See entry 104.

1200 *The Submarine Rescue Chamber: Description, Operation, Care and Upkeep*. Coco Solo, C. Z., Submarine Base, 1932.
Contents sheet plus 18 leaves. Paper covers.

Submarine Safety and Salvage. See entry 1394.

Submarine Safety, Respiration and Rescue Devices. See entries 1335, 1370.

Submarine Sailor. See entry 399.

1201 SUBMARINE SIGNAL COMPANY. *Submarine Signals: Record of Results Obtained by the Use of the System of Submarine Signaling controlled by the Submarine Signal Company*. Boston, 1906. 64 pp., illus. LC 22-5492.

1202 ________. *Submarine Signals: Results of Tests Made by the United States Lighthouse Board, during June and July, 1906, of the System of Submarine Signaling Controlled by the Submarine Signal Company*. Boston, The Heintzemann Press, 1906. 29 pp., tables. LC 6-40474.

1203 SUBMARINE SIGNAL COMPANY. *Submarine Signals. I. British Admiralty Tests of Submarine Signaling. II. List of Vessels Equipped. III. Opinions of British Ship Owners*. Boston, Submarine Signal Co., 1907. 22 pp. LC 7-9851.

1204 ________. *Submarine Signals for Safe Navigation*. Boston, Submarine Signal Corporation, 1925. 18 pp., illus. LC 25-11942.

Submarine Signaling. See entries 119, 588, 864.

Submarine signaling. See entries 10, 119, 393.

Submarine Signals. See entry 1304.

Submarine Stories and Adventures. See entry 1424.

Submarine Tankers. See entry 474.

Submarine Text. See entry 1395.

Submarine: The Autobiography of Simon Lake. See entry 740.

Submarine! The Story of Undersea Fighters. See entry 46.

The Submarine: The Ultimate Weapon. See entry 1454.

Subsurface Warfare. See entry 595.

Subs and Submariners. See entry 1453.

1205 SUETER, Murray Fraser, 1872- . *The Evolution of the Submarine Boat, Mine and Torpedo, From the Sixteenth Century to the Present Time*. Portsmouth, England, Griffin, 1907. xxiv, 384 pp., frontis., plates, diagrams. Glossary. LC 8-19282.

One of the classics of submarine literature.

Suicide Submarines. See entry 1487.

Summaries of Research Reported on During 1966. See entry 718.

The Sun Beneath the Sea. See entry 972.

Sunk: The Story of the Japanese Fleet. See entries 578, 579.

Sunken History. See entry 1133.

The Sunken Submarine. See entry 283.

The Super Submarine. . . . See entry 148.

Surcouf (French submarine). See entry 544.

Surface at the Pole. See entry 179.

Sur Mer. . . . See entry 206.

A Survey Report on Human Factors in Underseas Warfare. See entry 895.

Sutton, Eric. See entry 902.

1206 SVERDRUP, Harald Ulrik, 1888- . *Hvorledes of Hvorfor med Nautilus*. Oslo, Gyldendal Norsk Forlag, 1931. 183 pp., illus., plates, maps.

Re the Norwegian submarine Nautilus.

Swastika at Sea. See entry 86.

Sweden. See entry 520.

1207 SWEENEY, James B. *A Pictorial History of Oceanographic Submersibles*. London, R. Hale, 1972. 314 pp., illus. ISBN 0-7091-2864-9.

1208 ________. *A Pictorial History of Oceanographic Submersibles*. NY, Crown, 1971. 320 pp., with 420 illus.

The whole story from the weird Greek, Roman, and Persian contraptions to the more scientific achievements of the 17th century to the polaris-missile-equipped USS *George Washington*. Written by the Public Relations Officer of the US Naval Oceanographic Office.

Sweeney, James B. See also entry 780.

1209 SWEENEY, John. *Skin Diving and Exploring Underwater*. NY, McGraw-Hill, 1955. 176 pp., illus. LC 55-6870.

Swept Channels. See entry 332.

Swift, Tom. See entries 27, 28, 29.

Symposium on Modern Developments in Marine Sciences. See entry 15.

1210 SYMPOSIUM ON UNDERWATER PHYSIOLOGY, 1st. *Proceedings*. Edited by Loyal G. Goff. Washington, National Academy of Sciences, 1955. Illus.

1211 SYMPOSIUM ON UNDERWATER PHYSIOLOGY, 2nd. *Proceedings*. Edited by C. J. Lambertsen and L. J. Greenbaum, Jr. Washington, National Academy of Sciences, 1963.

1212 SYMPOSIUM ON UNDERWATER PHYSIOLOGY, 3rd. *Proceedings*. Edited by C. J. Lambertsen. Baltimore, Williams and Wilkins, 1967. 517 pp., illus. LC 66-28164.

This third symposium was held at the State Department in Washington on March 23-25, 1966 under the sponsorship of the Office of Naval Research, and the Committee on Undersea Warfare of the National Research Council.

Synthèse de la Guerre Sous-Marine. See entry 201.

T

Taboada, E. Nunez de. See entry 459.

Taffrail (pseud.) See Henry Taprell Dorling, entries 331, 332.

1213 TAILLEZ, Philippe, 1905- . *In Den Tiefen der Meere.* Translated by Herbert and Waltraut Furreg. Wien, Ullstein, 1956. 213 pp., illus. LC 56-27714.

1214 ________. *Plongées sans Cable.* Paris, Arthaud, 1954. 231 pp., illus. LC 54-23083.

1215 ________. *To Hidden Depths.* NY, Dutton, 1954. 188 pp., illus. LC 54-10157.

Tailliez, Philippe, 1905- . See also 425, 426, 631.

Take Her Down, A Submarine Portrait. See entries 1239, 1240.

1216 TALBOT, Frederick Arthur Ambrose, 1880- . *Submarines, Their Mechanism and Operation.* Philadelphia, Lippincott, 1915. London, Heinemann, 1915. 274 pp., frontis., illus., plates. LC A 16-747.

Tantum, W. H. See entry 341.

1217 TASSOS, John. *The Underwater World: A Complete Guide to Diving, Spearfishing, and Other Underwater Secrets.* Englewood Cliffs, N.J., Prentice-Hall, 1957. 242 pp., illus. LC 57-5557.

Tattoos. See entry 1352.

Tauchertechnik. See entry 1179.

Tauchkunst, der Moderne Sport. See entry 78.

1218 TAVOLGA, William Nicolai, 1922- , ed. *Marine Bio-Acoustics: Proceedings of a Symposium Held at the Lerner*

Marine Laboratory. Bimini, Bahamas. April 11-13, 1963. NY, Macmillan, 1964. 413 pp., illus. LC 63-23207.

Taylor, B. W. See entry 1121.

1219 TAYLOR, Joan Du Plat, ed. *Marine Archaeology: Developments During Sixty Years in the Mediterranean.* Edited for C.M.A.S. (World Underwater Foundation). London, Hutchinson, 1965. LC 65-9843. NY, Crowell, 1966. LC 65-20609. 208 pp., illus., maps.

1220 TAYLOR, Theodore, 1922- . *Fire on the Beaches.* NY, Norton, 1958. Toronto, McLeod, 1958. 248 pp., plates. LC 58-5151.

1221 TECHEL, H. *Der Bau von Unterseebooten auf der Germaniawerft.* Berlin, Verlag des Vereines Deutscher Ingenieure, 1922. 103 pp., illus., plates. LC 23-9296.

The Germaniawerft submarine shipyard in Kiel, Germany.

Teillac, Pierre, 1881- . See entry 1403.

1222 TEITCH, Clark. *A Modern Submarine Boat.* NY, N.W. Henley Pub. Co., 1912.

A 14″ x 28″ diagram of a submarine.

1223 TELFER, Dorothy Weaver. *Exploring the World of Oceanography.* Introduction by C. B. Hamn. Chicago, Children's Press, 1968. 87 pp., illus. LC 68-8975.

A juvenile work on submersible research craft.

Ten Miles High, Two Miles Deep. See entry 623.

Ten Years and Twenty Days. See entry 318.

The Tenth Fleet. See entry 392.

Teodoro. See entry 823.

Teoriia Podvodnykh Lodok. See entry 658.

Terrapin, USS (fictional). See entry 822.

Le Terreur des Mers. See entry 1403.

Il Terrore sui Mari. See entry 1404.

1224 TERRY, Richard D. *The Deep Submersible.* North Hollywood, Calif., Western Periodicals, 1966. 456 pp., illus. LC 66-6236.

Bathyscaphe and underwater exploration.

1225 ________, ed. *Ocean Engineering: A Preliminary Report* submitted to the Chairman of the Interagency Committee of Oceanography by the National Security Industrial Association. (Prepared by a staff at North American Aviation, Inc.) Published for the NSIA. North Hollywood, Calif., Western Periodicals, 1966. 4 volumes. LC 66-4187.

Testing. See entry 688.

1226 THEBERGE, James Daniel, ed. *Soviet Seapower in the Caribbean: Political and Strategic Implications.* NY, Praeger, 1972. 175 pp., maps. LC 72-75699.

1227 THEISS, Lewis Edwin, 1878- . *On Board a United States Submarine.* Boston, Wilde, 1940. 308 pp., plates. LC 40-34111.

Theories Strategiques. See entry 202.

Thetis, HMS. See entries 1426, 1428.

They Came to Fish. See entry 149.

They Dared the Deep. See entry 838.

1228 THIELEMANN, Walter. *U-Boot und Weltwirtschaft. Des Wirtschaftsleben im Gesamtunterricht der Arbeitsschule.* Prag, Haase, 1916. 279 pp., illus., plates, maps. LC 21-8138.

1229 THIES, Hans Arthur, 1893- . *Der Eiserne Seehund: Wilhelm Bauer, der Erfinder des U-Boots; Tat, Schicksal und Abenteurer.* Munich, Knorr & Hirth, 1943. 335 pp., plates, ports. LC AF 47-36.

1230 ________. *De Ijzeren Zeehond.* Translated by J. H. M. van der Eerden. Amsterdam, Uitgeverij Westland, 1943. 355 pp., frontis. port, plates. LC AF 47-5726.

This World and That. See entry 866.

1231 THOMAS, David Arthur, 1925- . *Submarine Victory: The Story of British Submarines in World War Two.* London, Kimber, 1961. 224 pp., plates, maps.

1232 THOMAS, Lowell Jackson, 1892- . *Raiders of the Deep.* 1st ed. Garden City, Doubleday, Doran & Co., 1928. 363 pp., illus.

1233 ________. *Raiders of the Deep.* NY, The Sun Dial Press, 1940. 363 pp., frontis., illus., facsims. LC 40-6244.

1234 ________. *Raiders of the Deep.* NY, Award, 1964. paperback reprint.

1235 ________. *Ritter der Tiefe: Unbersetzt und bearbeitet von E. freiherr v. Spiegel.* Berlin, Deutsche Verlagsgesellschaft, 1931. 366 pp., frontis., plates, ports, facsims. LC 32-17889.

1236 THOMASI, Alphonse. *Sous-Marins et Croiseurs Français.* Paris, Plon, 1947. 120 pp. LC A48-9420.

1237 THOMPSON, Frank E., Jr. *Diving, Cutting, and Welding in Underwater Salvage Operations.* Baltimore, Cornell Maritime Press, 1969, c. 1944. 214 pp., illus. SBN 87033-139-6. LC 70-92687.

1238 THOMPSON, Lawrence Roger, 1906- . *The Navy Hunts the CGR 3070.* Garden City, Doubleday, 1944. Toronto, McClelland, 1944. 150 pp., plates. LC 44-487.

1239 THOMPSON, Terry Brewster, 1891- . *Take Her Down, A Submarine Portrait.* NY, Sheridan House, 1937. 317 pp., plates. LC 38-27067.

WW I submarine L-9.

1240 ________. *Take Her Down, Wartime Adventures of U.S. Submarine L-9.* NY, Sheridan House, 1942. 313 pp., plates. LC 42-13430.

1241 THOMS, Herbert, 1885- . *Doctors of Yale College, 1702-1815: and the Founding of the Medical Institution.* (Yale University School of Medicine. Department of the History of Medicine. Pub. #39) Hamden, Conn., Shoe String Press, 1960. xxi, 199 pp., illus. LC 60-15841.

Information on David Bushnell and Revolutionary War submarine Turtle.

1242 THOMSON, George Pirie, 1887- . *Submarines.* Illustrated by K. E. Carter. London, Muller, 1959. New Rochelle, Soccer, 1959. 144 pp., illus., diagrams. LC 59-3223.

Detailed description of the working of a submarine.

1243 THORNE, Guy. *The Secret Service Submarine.* NY, Sully and Kleintech, 1915. 190 pp., illus.

1244 THORNE, Jim. *Adventures Under the Sea.* NY, Walker, 1965. Toronto, McLeod, 1965. 187 pp., illus. LC 65-15990.

Scuba and skin diving for juvenile audience.

1245 ________. *Jim Thorne's Guide to Adventure.* NY, Crowell, 1967. 192 pp., illus. LC 67-18527.

A travel itinerary for skin divers.

1246 ________. *The Underwater World: A Survey of Oceanography Today.* NY, Crowell, 1969. 160 pp., illus. LC 69-18673.

1247 ________. *Underwater World.* NY, Barnes and Noble, 1970.

Underwater exploration.

Three Boys on an Electrical Boat. See entry 1267.

300,000 Tonnen Versenkt! See entry 1405.

1248 THRESHER (SSN 593). *United States Ship Thresher. In Memoriam April 10, 1963.* Washington, U.S. Government Printing Office, 1964. 146 pp., illus., ports. (D 201.2:T 41)

Thresher, USS (SSN 593). See also 989, 1146.

The Thresher Disaster. See entry 96.

Threshold: A Novel. See entry 257.

1249 THROCKMORTON, Peter. *The Lost Ships: An Adventure in Undersea Archaeology.* Boston, Little, Brown, 1964. Toronto, 1964. 260 pp., plates. LC 64-15049.

Discovery of Bronze Age ships.

1250 ________. *Shipwrecks and Archaeology: The Unharvested Sea.* Boston, Atlantic Monthly Press, 1970. 270 pp., illus. LC 76-79373.

Through Hell and Deep Water. See entry 793.

Thule, HMS. See entry 834.

Thunderbolt, HMS. See entry 1428.

A Time of Killing. See entries 565, 566.

1251 TINKER, Gene. *The Skin Diver's Travel Guide: Pan Am's Guide to Snorkeling and Scuba-diving in Bermuda, the Bahamas, the Caribbean, and the Mediterranean.* NY, Doubleday, 1967. 336 pp., illus. LC 67-19036.

To Hidden Depths. See entry 1215.

Der Tod auf Allen Meeren. See entry 444.

1252 TOGO, Kichitaro. *The Naval Battles of the Russo-Japanese War.* Translated by J. Takakusu. Tokyo, 1970.

Tom Swift and his Jetmarine. See entry 28.

Tom Swift and his Submarine Boat. See entry 27.

Tom Swift and his Subocean Geotron. See entry 29.

1253 TOMASHEVICH, A. V. *Podvodnye Lodki v Operatsiiakh Russkogo Flota no Baltiiskom More v 1914-1915 gg.* (Submarines in the operations of the Russian Fleet in the Baltic during 1914-1915.) Moscow, 1939- . Multi-volumed work. LC 42-45241.

1254 TOMASSI, Gaetano Marchetti. *Nuovo Trattato sulla vera rettificazione del Circolo. Con una Dissertazione sul modo di navigare sott'acqua invento dalla stesso Autore l'anno 1799.* Edizione seconda. Fuligno, Feliciano Campitelli, 1817. 75 pp., 34 pp., 1 blank leaf. With 6 folding copperplate illus.

1255 [TOMLINSON, E., 1859-1931]. *David Bushnell and his American Turtle.* NY, Akron, Chicago, The Werner Co., 1899. 47 pp., illus.

This work has little bearing on the facts.

1256 TOMMASI, Donato, 1848-1907. *Les Bateaux Hémi-Plongeurs: Nouveau Type de Construction Navale Applicable Soit à la Marine Marchande, Soit à la Marine de Guerre.* Paris, Walder, 1876. 40 pp., plates. LC 30-31046.

Tomorrow Never Came. See entry 203.

Torpeda i Mina. See entry 761.

Torpedo Achtung! See entry 837.

The Torpedo in Peace and War. See entry 677.

Torpedo-und Minenkrieg. See entry 1080.

Torpedo War and Submarine Explosions. . . . See entries 455-458.

Torpedoes. See entries 6, 32, 49, 137, 195, 270, 286, 322, 327, 360, 452, 453, 455-459, 476, 558, 611, 677, 735, 755, 761, 921, 961, 1080, 1144, 1177, 1190, 1192, 1205, 1329, 1378.

Torpedoes and Torpedo-Vessels. See entry 32.

Torpedoes and Torpedo Warfare. See entry 1141.

Torpedoes Away! See entries 936, 1129.

Torpedoes Away, Sir! See entry 584.

Torpedoes, Mark 15 Type. See entry 1366.

Les Torpilles et les Mines Sous-Marines. See entry 921.

Torpilles et Torpilleurs de Nations Etrangeres. See entry 158.

Torpilles Humaines. See entry 943.

Torpilleurs et Sous-Marins. See entry 994.

1257 TOSCHI, Elias. *Ninth Time Lucky.* Translated by James Cleugh. London, Kimber, 1955. 216 pp., illus.
Italian midget submarines. WW II.

1258 TOUDOUZE, George Gustave, 1877- . *Le Sous-Marin, Roi de la Mer.* Paris, Lemerre, 1920. 266 pp., plates.

1259 ________. *Le Sous-Marin, Roi de la Mer.* Paris, Lemerre, 1949.

Tournus, G. de. See entry 44.

Towing. See entry 1377.

1260 TOWNSEND, Derek. *Underwater Photography; Movies and Still.* London, Paul, n.d. 160 pp., illus., plates.

The Tragedy of the Lusitania. See entry 366.

Training. See entries 769, 1343, 1345.

Traité de Plongée. See entry 547.

Traits . . . Related to Submariner Adjustment. See entry 1347.

1261 TRAPP, Georg von. *Bis Zum Letzten Flaggenschuss: Erinnerungen Eines Österreichischen U-Boots-Kommandanten.* Salzburg, Leipzig, Pustet, 1935. 247 pp., frontis., illus., maps, plates, ports. LC 36-10360.
Memoirs of a WW I Austrian submarine commander who fled Europe for America and founded the Trapp Family Singers.

Travels in England. See entry 979.

1262 TRAVINICHEV, Alessei Platonovich. *Ocherki po Borbe s*

Podvodnymi Lodkami. Moscow, 1938. 197 pp., illus., maps, plans, diagrams, bibliography. LC 49-33810.

Russian account of WW I submarine warfare.

Treasure Diving Holidays. See entry 275.

Treasures Beneath the Sea. See 1134.

Treub, M.W.F. See entry 624.

1263 TRÉVILY, Joseph. *Essai sur L'Histoire et la Situation du Sous-Marin en Droit International*. Paris, Librairie du Recueil Sirey, 1931. xiii, 251 pp., bibliography. LC 32-21516.

Submarine warfare considered from standpoint of international law.

1264 TREW, Antony. *Two Hours to Darkness*. NY, Random House, 1963. 312 pp., illus. LC 63-7124.

A novel in the vein of *Fail Safe* and *Seven Days in May*. An it-could-happen-here kind of story in which the commander of a nuclear submarine, cracks up and tricks crew into believing they must act to stop an impending attack of the Russians.

1265 TRIESTE. *Welcome Aboard. Bathyscaph Trieste*. San Diego, U.S. Navy Electronics Laboratory, 1959. 15 pp., illus.

A public relations pamphlet which describes workings of bathyscaphe with words and pictures.

The Trieste. See entry 1146.

Trieste (bathyscaphe). See entries 825, 971, 1146, 1152, 1265.

1266 TRIMBLE, Hugh J., 1924- . *Return from the Deep*. Indianapolis, McHew, 1958. 197 pp. LC 59-16077.

Quasi-fictional story of the sinking and raising of the USS Squalus, and her subsequent career during WW II as the USS Sailfish. Also employs use of the soubriquet "Squailfish."

Tripoli. See entry 307.

Triton (diving dress). See entry 340.

Triton, USS (SSRN 586). See entries 71, 75, 304, 305.

1267 TROWBRIDGE, John. *Three Boys on an Electrical Boat*. 6th ed. Boston, Houghton, Mifflin, 1896. (c. 1894). 215 pp.

Juvenile fiction.

True Book About Submarines. See entry 554.

1268 TRUMBULL, Robert Oliver, 1912- . *Silversides.* NY, Holt, 1945. Toronto, Oxford, 1945. 217 pp., plates, ports. LC 45-7556.

Story of WW II submarine USS Silversides (SS 236).

1269 TRUSOV, G. M. *Podvodnye Lodki v Russkom i Sovetskom Flote.* (Submarines in the Russian and Soviet Fleet). Moscow, 1957, Leningrad, 1963. 440 pp., illus. LC 64-44414.

1270 TSIPIS, Kosta, ed. *The Future of the Sea-Based Deterrent.* Cambridge, MIT Press, 1973. 266 pp. ISBN 0-262-56012-7. LC 73-16067.

Papers of a symposium held in Racine, Wisconsin November 1972, organized by the Carnegie Endowment for International Peace and the American Academy of Arts and Sciences under the Sponsorship of the Johnson Foundatin and the Alfred P. Sloan Foundation.

1271 TUCKER, Gilbert Norman, 1898- . *The Naval Service of Canada: Its Official History.* Ottawa, King's Printer (Published under the authority of the Minister of National Defence) 1952. 2 volumes, illus., ports, maps. LC A 53-856.

There are references to Canadian submarines scattered throughout the two volumes. The principal references are in Volume I, Chapter 13.

Türk Denizalticilik Tarihi. See entry 855.

Turkey. See entries 1300, 1311, 1319.

Turkish prisons. See entry 1189.

Turkish submarines. See entry 855.

1272 TURNER, John Frayn. *Periscope Patrol: The Sage of Malta Submarines.* London, Harrap, 1957. Toronto, Clarke, Irwin, 1957. 218 pp., plates.

Reissued as a 160 pp. paperback in the "Panther" edition. London, Hamilton, 1958.

Turner, Richmond Kelly. See entry 389.

Turtle (Bushnell's). See entries 2, 3, 617, 633, 1241, 1255.

Tusk, USS (SS 426). See entry 762.

Twenty-First Century Submarine. See entry 593.

Twenty Million Tons Under the Sea. See entry 467.

Twenty Thousand Leagues Under the Sea. See entry 1409.

U

U-Boat Warfare. See entry 548.

U-Boats at War. See entries 167, 168, 169.

U-Boats Destroyed. See entry 505.

U-Boats to the Rescue. See entry 956.

U-Boats Westward! See entry 575.

U-Boot-Abenteuer. See entry 1162.

U-Boot Abenteuer im Operrgebiet. See entry 471.

U-Boot-Abenteuer im Sperrgebiet. See entry 1114.

Das U-Boot als Kriegs-und Handelsschiff. See entry 733.

Das U-Boot bei der Arbeit. See entry 707.

1273 *U-Boot-Brevier: Ein Buch von Unseren Helden Zur See.* Sketches by Hans Aburi, S. Baske, Max Bewer, and others. Pictures by H. von Bartels, K. Becker, J. Bahr, and others. Berlin, Eysler, 1917. 159 pp., illus. LC 21-20229.

A song "U9" with music by W. Espe appears on pp. 124-127.

U-Boot, Englands Tod! See entry 1178.

U-Boot-Fahrer und Kamelsreiter. See entry 307.

U-Boot-Faher von Heute. . . . See entry 1103.

U-Boot gegen U-Boot. See entry 591.

U-Boot und Luftboot. See entry 706.

U-Boot und U-Boot-Krieg. See entry 1421.

U-Boot und Weltwirtschaft. See entry 1228.

U-Boot vor dem Feind. See entry 419.

Der U-Boot-Krieg. See entry 634.

Der U-Bootkrieg Als Weg zum Endsieg. See entry 39.

Der U-Boot-Krieg des Jahres 1915. See entry 461.

U-Boote. See entries 1061, 1108.

U-Boote am Feind. See entry 748.

U-Boote, Deutschlands Scharfe Waffe. See entry 492.

U-Boote gegen England. See entry 826.

U-Boote im Einsatz 1939-45. See entry 604.

1274 *U-Boote im Eismeer.* von ** . Berlin, Scherl, 1917. 102 pp., map. LC 34-6235.

U-Boote vor New-York. See entry 80.

U-Boote Westwärts! See entries 576, 577.

U Bootsarbeit und ihr Erfolg. See entry 954.

Die U-Bootserfolge der Achsenmachte 1939-1945. See entry 1062.

U-Bootsfahrten. See entry 164.

U-Bootsfallen. See entry 1023.

Ubootsgeist. See entry 1038.

U-Boots-Kommandanten und Kriegsverbrecher. See entry 420.

Der U-Bootskrieg, 1914-1918. See entry 857.

U-Boots-Leben. See entry 421.

U-Boots-Maschinist Fritz Kasten. See entry 441.

U-Boots-Taten. See entry 165.

Die U-Bootswaffe. See entry 318.

U-Boottyp XXI. See entry 1054.

U-Boottyp XXIIII. See entry 1055.

U-Deutschlands Fahrt. See entry 1139.

Udar Pod Vodoi. See entry 961.

1275 UHRE, Jan. *Frømandsfaerd: Den Moderne Dyknings Logbog.* Copenhagen, Hassing, 1956. 157 pp., illus. LC 57-36310.

U-Kreuzer 151 Greift An. See entry 978.

Unbroken: The Story of a Submarine. See entry 836.

Under Sea with Helmet and Camera. See entry 353.

Under the Black Ensign. See entry 549.

Underwater Acoustics. See entry 662.

Underwater acoustics. See entry 102.

Underwater Archaeology. See entry 973.

1276 *Underwater Archaeology.* NY, UNESCO, Unipub, 1972. 340 pp.

Underwater archaeology. See entries 58, 59, 60, 116, 170, 245, 294, 308-313, 333, 334, 350, 387, 388, 435, 445, 694, 782, 820, 838, 966, 967, 1007, 1133, 1219, 1249, 1250, 1462.

Underwater Defense Handbook. See entry 178.

Underwater Demolition Teams (UDT). See entries 389, 390, 391, 436, 538, 609, 1002, 1003, 1193, 1422.

Underwater Diving Manual. See entry 150.

Under-Water Exploration. See entry 313.

Underwater Exploration. See entries 935, 1153.

Underwater Explosions. See entry 238.

The Underwater Guide to Marine Life. See entry 1014.

The Underwater Handbook. See entry 162.

Underwater Medicine. See entries 861, 862, 863.

1277 *Underwater Medicine and Related Sciences: A Guide to the Literature.* Charles W. Shilling and Margaret F. Werts, editors. NY, Plenum Press, 1973. 695 pp. ISBN 0-306-65161-0.

Underwater Photography. See entry 1260.

Underwater Photography. See entry 1172.

Underwater Physiology. See entries 1210-1212.

Underwater Research. See entry 783.

Underwater Research Vehicles. See entry 1223.

Underwater Saboteur. See entry 827.

Underwater Safety. . . . See entry 279.

Underwater Ships of Science, USSR. See entry 1079.

Underwater Society of America. See entry 348.

Underwater Sport. See entry 1407.

Underwater: The Skindiver's Manual. See entry 48.

Under-Water Torpedo-Boats. See entry 745.

The Underwater War. Submarines, 1914-1918. See entry 509.

Underwater Warriors. . . . See entry 105a.

Underwater Welding and Cutting. See entry 708.

Underwater Work. See entry 204.

Underwater Work. See entry 1237.

The Underwater World. See entries 1217, 1246, 1247.

Underwater World: Exploration Under the Surface of the Sea. See entry 236.

1278 *Unterseeboote und Seeminen.* Leipzig, Paul, 19--. 38 pp.

Note: This entry is out of alphabetical sequence, but is retained here for cross-reference purposes.

1279 U.S. ATOMIC ENERGY COMMITTEE. *Naval Reactor Program and Polaris Missile System.* Hearing, 86th Congress, 2nd Session, on review of progress in the naval reactor program and developments in the Polaris missile submarine system. April 9, 1960. Washington, 1961. vii + 39 pp., illus. LC 61-61449. (Y4.At 7/2: R 22/12)

1280 ________. *Report of Underseas Warfare Advisory Panel to Subcommittee on Military Applications.* August 1958. v. + 16 pp. (85th Congress, 2nd Session. Joint Committee print) LC 58-61936.

1281 ________. *Review of Naval Reactor Program and Admiral Rickover Award.* Hearings, 86th Congress, 1st Session. April 11 and 15, 1959. Washington, 1959. vi + 81 pp. LC 59-61940.

1282 ________. *Submarine Advanced Reactor Gas Turbine Cycle Study* by E. A. Sheehan and A. O. White. September 29, 1953. Schenectady, NY, General Electric Company, General Engineering Laboratory, 1955. (2) + v + 70 pp., illus. LC 55-60899.

1283 U.S. COAST ARTILLERY OFFICE. *Submarine Mining.* Washington, USGPO, 1930. 275 pp., illus., tables. (War Dept. Training Manual No. 2160-20) LC 31-26224.

1284 U.S. 64th CONGRESS, 1st SESSION. 1915-1916. *Peace or War? The Great Debate in Congress on the Submarine and the Merchantman.* Compiled from the *Congressional Record* by William Bayard Hale. n.p., The Organization of American Women for Strict Neutrality, 1916. 155 pp. LC 17-17995.

1285 U.S. CONGRESS. HOUSE ARMED SERVICES COMMITTEE. *Subcommittee Hearing on H.R. 1227 to Amend Further Act to Authorize Construction of Experimental Submarines.* . . . Approved May 16, 1947 as amended. (Paper # 32) Washington, 1951. pp. 721-723. (82nd Congress)

1286 ________. *Full Committee Hearing on H.R. 5956 to Authorize Loan of Two Submarines to the Government of Italy.* (83rd Congress. House. ASC Paper # 33, pp. 1379-91). Washington, 1953.

1287 U.S. CONGRESS. HOUSE. COMMITTEE ON NAVAL AFFAIRS. *Fleet Submarines.* To accompany H.R. 13670. Washington, 1916. 5 pp. (64th Congress, 1st Session. House. Report 492 submitted by Lemuel Phillips Padgett). LC 16-26333.

1288 ________. *Hearings before the Committee on Naval Affairs . . . on Submarine Boats.* Washington, 1902. 314 pp., illus., plates. LC 5-5032.

Lake, Holland and Moriarity submarine boats. Statements by company representatives and by naval representatives.

1289 ________. *Hearings on H.R. 19617 and H.R. 19618, providing for an Increase of the Navy by Purchase of Ten Submarine Boats, and So Forth. February 4, 1910.* Washington, 1910. 19 pp. LC 10-35192.

1290 ________. *Storage Batteries for Submarines.* Extracts from the hearings of Hon. Josephus Daniels, Secretary of the Navy, . . . December 11, 1916. Washington, 1917. LC 17-26347.

1291 U.S. 82nd CONGRESS. HOUSE REPORTS. # 550. *Amending Further the Act to Authorize Construction of Experimental Submarines and for Other Purposes.* Approved May 16, 1947 as amended. Report from the Committee on Armed Services to accompany H.R. 1227. June 12, 1951. Washington, 1951. 4 pp.

1292 ———. #773. *Consideration of H.R. 1180. Report from Committee on Rules to Accompany House Resolution 858. July 30, 1951.* Washington, 1951. 1 page.

1293 U.S. 83rd CONGRESS. HOUSE REPORTS. #756. *Authorizing Loan of two Submarines to the Government of Italy.* . . . Report from Committee on Armed Services to accompany S.2277. July 9, 1953. Washington, 1953. 4 pp.

1294 U.S. 84th CONGRESS. HOUSE REPORTS ON PUBLIC BILLS. #1687. *Authorizing Loan of Two Submarines to the Government of Brazil.* Report from the Committee on Armed Services to accompany H.R. 8100. January 30, 1956. Washington, 1956. 4 pp.

1295 U.S. 86th CONGRESS. HOUSE. ARMED SERVICES COMMITTEE. #52. *Subcommittee #1 Consideration of H.R. 10500, to Amend Career Compensation Act of 1949 with Respect to Incentive Pay for Certain Submarine Service.* Washington, 1960. pp. 4013-4025.

1296 ———. #1570. *Retaining Entitlement to Incentive Pay for Personnel Assigned to Operation of Nuclear Submarines.* Report from Committee on Armed Services to Accompany H.R. 10500. May 3, 1960. Washington, 1960.

1297 U.S. CONGRESS. HOUSE. SELECT COMMITTEE UNDER HOUSE RESOLUTION 288. *Hearings beginning March 9, 1902-(April 30, 1908).* Washington, 1908. 2 vols. facsims. LC 8-35502.

Investigation of charges by George L. Lilley against the Electric Boat Company.

1298 U.S. CONGRESS. PUBLIC LAWS. #176. *H.R. 1227 Act to Amend Further the Act to authorize Construction of Experimental Submarines, and for Other Purposes.* Approved May 16, 1947, as amended. Approved October 16, 1951. 1 pp. Chapter 506. 1st Session.

1299 U.S. 83rd CONGRESS. Public Law 188. *S. 2277. Act to Authorize Loan of Two Submarines to the Government of Italy.* . . . Approved August 5, 1953. 1 pp. Chapter 321. 1st Session.

1300 U.S. 83rd CONGRESS. PUBLIC LAWS. #214. *S.2539 Act to Authorize Loan of Two Submarines to the Government of*

Turkey. Approved August 7, 1953. 1 pp. Chapter 347. 1st Session.

1301 U.S. 84th CONGRESS. PUBLIC LAWS. #484. *H.R. 8100 Act to Authorize Loan of Two Submarines to the Government of Brazil*. Approved April 6, 1956. 1 pp. Chapter 187. 2nd Session.

1302 U.S. 82nd CONGRESS. PUBLIC LAWS. #510. *S.3337. Act to Authorize Loan of Two Submarines to the Government of the Netherlands*. Approved July 11, 1952. 1 pp. Chapter 675. 2nd Session.

1303 U.S. CONGRESS. SENATE ARMED SERVICES COMMITTEE. 31st Report. *Report on Conditions at Six Navy and Marine Corps Training Centers, Submarine Base, New London, Conn. . . . 1951*. v + 18 pp. (Committee Print 82nd Congress 1st Session) LC 50-61319.

1304 U.S. CONGRESS. SENATE. COMMITTEE ON COMMERCE. *Submarine Signals*. Report to accompany S.6603. Washington, 1910. 8 pp. (61st Congress. 2nd Session. Senate Report 326) LC 10-35309.

1305 U.S. CONGRESS. SENATE. COMMITTEE ON MILITARY AFFAIRS. *Cargo Aircraft and Cargo Submarines*. Hearings before a subcommittee of the Committee on Military Affairs. U.S. Senate. 77th Congress, 2nd Session. On S.J. Resolution 156, a joint resolution to provide additional transportation facilities for furnishing supplies to the armed forces of the U.S. in foreign countries. . . . Washington, 1942. 131 pp. LC 42-25495.

1306 U.S. CONGRESS. SENATE. COMMITTEE ON NAVAL AFFAIRS. *Fleet Submarines*. Report to accompany H.R. 13670. Washington, 1916. 4 pp. (64th Congress. 1st Session. Senate. Report 526.) LC 16-26576.

1307 ________. *Staff Corps Increase. Submarine Boats*. Hearings . . . May 29 and May 31, 1902. Washington, 1902. 103 pp. (57th Congress. 1st Session. Senate. Doc. 395.) LC 2-23230.

1308 ________. *Investigation of the Sinking of the Submarine S-4*. Hearings before the subcommittee of the Committee on Naval Affairs. U.S. Senate. 70th Congress 1st Session. Pursuant to Senate resolution 205. . . . Washington, 1928. 326 pp.

1309 U.S. CONGRESS (82nd) SENATE REPORTS. #751. *Amending Further the Act to Authorize Construction of Experimental Submarines* . . . Approved May 16, 1947. . . . Report from Committee on Armed Services to accompany H.R. 1227. Sept. 14, 1951. 3 pp.

1310 U.S. CONGRESS (83rd) SENATE REPORTS. #497. *Authorizing Loan of Certain Naval Vessels to Certain Foreign Governments* . . . (Italy) to accompany S. 2277. July 2, 1953. Washington, 1953. 8 pp.

1311 ________. *#823. Loan of Two Submarines to the Government of Turkey.* Report from the Committee on Armed Services to accompany S.2539. July 31, 1953. Washington, 1953. 2 pp.

1312 U.S. CONGRESS (84th) REPORT ON PUBLIC BILLS. *Authorizing Loan of Two Submarines to the Government of Brazil.* Report from Committee on Armed Services to accompany H.R. 8100. March 22, 1956. Washington, 1956. 4 pp.

1313 U.S. CONGRESS (86th) REPORTS ON PUBLIC BILLS. #1298. *Authorizing Loan of One Submarine to Canada and Extension of Loan of Naval Vessel to Government of Republic of China.* Report from Committee on Armed Services to accompany H.R. 9465. April 28, 1960. Washington, 1960. 4 pp.

1314 U.S. CONGRESS. SENATE. SPECIAL COMMITTEE TO INVESTIGATE THE MUNITIONS INDUSTRY. *Munitions Industry.* Hearings before the Special committee investigating the munitions industry, U.S. Senate, 73rd and 74th Congresses. pursuant to S. Res. 206. . . . Washington, 1934-43. 40 volumes. illus., maps, tables, diagrams. LC 38-4.

Electric Boat Co. was one of the firms under investigation.

1315 U.S. DEPARTMENT OF STATE. *Diplomatic Correspondence Between the United States and Belligerent Governments Relating to Neutral Rights and Commerce.* . . . NY, Oxford University Press, 1917. 393 pp. LC 19-13164.

1316 ________. *Papers Relating to Maritime Danger Zones and the Placing by Belligerents of Mines in the High Seas.* Washington, 1917. 34 pp. LC 17-12814.

1317 U.S. STATE DEPARTMENT. TREATIES & OTHER INTERNATIONAL ACTS series. #3124. Mutual Defense Assistance. Loan of Submarines to Italy, Understanding between the United States and Italy, effected by exchange of notes, Signed Washington April 27, 1954. . . . Washington, 1955. (pub. #5439) (2) + 4 pp. LC 55-63381.

1318 ________. #3731. Mutual Defense Assistance. Loan of Vessels to Brazil, agreement between the United States and Brazil, effected by exchange of notes, signed Washington January 12 and 16, 1957. Washington, 1957. (2) + 5 pp., text in English and Portuguese. LC 57-61128.

This agreement involved the fleet submarines *Muskallunge* and *Paddle*.

1319 ________. #8042. Loan of Submarines to Turkey, agreement between the United States and Turkey, effected by exchange of notes, signed Ankara, February 16 and July 1, 1954. Washington, 1955. (Pub. #5650) (2)+4 pp. LC 55-63462.

1320 ________. Mutual Defense Assistance. Loan of Submarines to Italy, agreement between the United States and Italy, amending agreement of April 27, 1954, effected by exhange of notes, signed Rome January 29, 1960. Washington, 1960. (2)+3 pp. Text in English and Italian. LC 60-61157.

1321 U.S. DISTRICT COURT. NEW YORK (SOUTHERN DISTRICT). *In the Matter of the Petition of the Cunard Steamship Company, Limited, as Owners of the Steamship "Lusitania" for Limitation of its Liability*. Before the Hon. Julius M. Mayer, District Judge. Opinion of the Court. NY, 1918. 45 pp. LC 18-22718.

1322 U.S. JOINT PUBLICATIONS RESEARCH SERVICE. SERIES. #6273. *Selected Soviet Military Translations:* No. 42. *Combat Training of Submarine Crews*. by P. Sinetskiy. Nov. 23, 1960. 15 pp. LC 59-61686.

1323 U.S. LIBRARY OF CONGRESS. Catalog of OSRD Reports. Division 6. *Subsurface Warfare*. (Unclassified Reports). Compiled and edited by Sylvia E. Weiss and Elizabeth C. McCue. August 1951. (63)+106 pp. (Navy Research Section) LC 50-60310.

1324 U.S. NAVAL CONSULTING BOARD. *The Enemy Submarine*. (Bulletin no. 2. May 1, 1918) NY, 1918. 47 pp. LC 18-26725.

1325 ________. *The Submarine and Kindred Problems.* NY, 1917.

1326 U.S. NAVY. NAVAL HISTORY DIVISION. *Partial Checklist of World War II Histories and Historical Reports in the U.S. Naval History Division.* Washington, 1973.

1327 U.S. NAVAL OCEANOGRAPHIC OFFICE. *Manned Submersibles and Underwater Surveying.* Washington, 1970. (Superintendent of Documents Pub.#SP-153)

1328 *U.S. Naval Submarine Base. New London, Connecticut.* New London, ca. 1959. 36 pp. illus.

General information pamphlet concerning facilities and activities of the base. Useful to Naval personnel and to visitors.

1329 U.S. NAVAL UNDERWATER ORDNANCE STATION, NEWPORT, R.I. *Handbook for Seaman Gunners. Catechism of Whitehead Torpedoes, Naval Defence Mine, Countermine, Observation Mine, Explosives, and Electricity.* Washington, GPO, 1908. 2178 pp. LC 8-35228 rev.

1330 U.S. NAVAL WAR COLLEGE. *Employment of Submarines.* Newport, ca. 1938.

1331 ________. *Submarine Warfare.* Newport, ca. 1938.

1332 U.S.N. BUREAU OF CONSTRUCTION AND REPAIR. *Diving Manual, 1924.* (Reprint of Chapter 36 of the Manual of the Bureau . . .) Washington, 1925. 108 pp., illus., tables, plates, plans, formulae. LC 25-26830 revised.

1333 ________. *Large Cast Steel Hooks . . . Submarine Lifting Pad Eyes and Test of Same. . . .* Washington, 1929. (Technical Bulletin No. 2-29) 6 pp., illus., tables. LC 30-2865.

1334 ________. *Report on Deep Diving Tests* by Gunner, G. D. Stillson. Washington, 1915. 252 pp., plates, diagrams. LC 15-26464.

Full name: George David Stillson, 1882-

1335 ________. *Submarine Safety, Respiration and Rescue Devices.* Washington, 1938. 137 pp., illus., plates. LC 38-26687.

1336 U.S.N. BUREAU OF ENGINEERING. *Lead Acid Storage Batteries. Instructions for the Installation, Operation and*

Care in Submarines. Prepared by F. A. Daubin. Washington, 1917. 153 pp., illus. LC 50-46176.

1337 U.S.N. BUREAU OF MEDICINE AND SURGERY. *Submarine Medicine Practice*. (NavMed-P 5054) (NavPers 10838-A) Washington, 1956. 357 pp., illus., decompression tables. LC 56-60989.

1338 ________. MEDICAL RESEARCH LABORATORY REPORTS #184. *Manual of Free Escape from Submarines.* by Harry J. Alvis. (Porject NM 002 015.01) 1952. 11 leaves. (volume 11)

1339 ________. ________. #185. *Theoretical Considerations of Use of Air-Filled Submarine Escape Appliance from Great Depths.* by Harry J. Alvis. (Project NM 002 015.08.01) vol. 11. 1952. 16 leaves, plate.

1340 ________. ________. #243. *Study of Thematic Apperception Test Stories and Sentence Completions of Subjects in Operation Hideout.* by Leonard D. Eron. and Frank Auld, Jr. Feb. 4, 1954. iii + 64 pp. v. 13 #4 Project NM 002-015.11.02.

1341 ________. ________. #252. *Visual Characteristics of Submarine Population.* by Ira Schwartz and N. Elaine Sandberg. June 21, 1954. iii+12+1 pp., illus. v. 13 # 12. Project NM 003041.57.02.

1342 ________. ________ #253. *Effect of Time in Submarine Service on Vision.* by Ira Schwartz and N. Elaine Sandberg. August 30, 1954. iii+9+1 pp., illus. v. 13 #14. Project NM 003 041.57.03.

1343 ________. ________. #259. *Predicting Success in Submarine School.* by Benjamin B. Weybrew. Nov. 2, 1954. v+37+2 pp., illus. v. 13 #20. Project NM 003 041.53.02.

1344 ________. ________. #262. *Claustrophobic Reactions to Some Stresses of Submarine Service.* by Jack L. Kinsey and Henry B. Murphree. April 4, 1955. 1+v leaves + 39+2 pp., illus. v. 14 #2. Project NM 008 041.53.03.

1345 . #263. *Prediction of Submarine School Success Through Use of Selected Themes in Navy Thematic Apperception Test.* by Robert J. Foster and Jean G. Carnaghan. May 2. 1955. iii+10+2 pp. illus., plate. v. 14 #3. Project NM 003 0451.54.01.

1346 ________. ________. #264. *Submarine Escape Training Experience, Summary of 25 Year Period. July 1929-December 1954.* with bibliog. by J. H. Peirano, H. J. Alvis, G. J. Duffner. May 12, 1955. iii, +28+2 pp. v. 14 #4. Project NM 002 015.08.03.

1347 ________. #279. *Exploratory Study Designed to Suggest Clusters of Traits and Assessment Tests Related to Submariner Adjustment.* by Benjamin B. Weybrew. Jan. 9, 1957. (1) + iii+14+1 pp., illus. v. 16#1 Project NM 003 041.54.03.

1348 ________. #301. *Relationship Between Submarine School Performance and Scores on Navy Thematic Apperception Tests.* by Bert T. King. October 1, 1958. 1+ii+15+1 pp. v. 17#8 Project NM 2302 20.01.04.

1349 ________. #305. *Exploratory Study of Efficiency of Dioctyl Sodium Sulfosuccinate (Colace) in Control of Constipation of Submariners During Prolonged Snorkel Cruise.* by Benjamin B. Weybrew, Richard A. Elias, and Joseph Vogel. Dec. 15, 1958. iv +19+1 pp. v. 17 #12. Project NM 24 01 20.04.04.

1350 ________. #307. *Experimental Study of Relationship of Autonomic Resiliency to Manifest Anxiety and Selected Personality Traits.* by Benjamin B. Weybrew and D. Alves. March 3, 1959. iii leaves + 44 pp. illus. v. 18 #2 Project NM 23 02 20.01.05.

1351 ________. #313. *Predicting Submarine School Attrition from Minnesota Multiphasic Personality Inventory.* by Bert T. King. August 20, 1959. iii + 25 pp. v. 18 #8 Project NM 23 02 20.01.06.

1352 ________. #319. *Relationship of Tattoos to Personal Adjustment Among Enlisted Submarine School Volunteers.* by Richard P. Youniss. Sept. 17, 1959. iii+10+1 pp. v. 18 #14 Project NM 23 02 20.01.07.

1353 ________. #321. *Approaches to Study of Motivation of Officer Candidates for Submarine Service.* by B. B. Weybrew and H. B. Molish. October 16, 1959. iv+47 pp., illus. v. 18 #16 Project MR 005 14 2100.02.02.

1354 USN BUREAU OF NAVAL PERSONNEL. *Facts About the Submarine Service.* Washington, 1944. 32 pp., illus., ports. (NAVPERS- NRB-44205) LC 47-23154.

1355 ________. *The Fleet Type Submarine.* 204 pp., illus., diagrams, plus 28 folding plates. (NavPers 16160. June 1946)

A training manual prepared for ComSubLant.

1356 ________. *Introduction to Sonar.* Prepared by Navy Training Publication Center. Washington, 1963. 162 pp., illus., diagrams. (NavPers 10130-A)

1357 ________. *Mineman 3c & 2c.* Prepared by Standards and Curriculum Division, Training, Washington, 1946 (1949). 357 pp., illus. LC 56-18525.

1358 ________. *Principles of Naval Ordnance and Gunnery.* Washington, 1959. 493 pp., illus., diagrams. (NavPers 10783)

1359 ________. *Submarine.* Prepared by Submarine School, Submarine Base. New London, Conn. (NavPers 16160) Revised May 1955. Published 1956. viii+197 pp., illus., 4 plates. LC 56-60755.

Issued in loose-leaf form. a revision of *Fleet Type Submarine.*

1360 ________. *Submarine.* Revised May 1955. Change 1, 1958. 24 pp., illus., plate. (NavPers 16160-A)

1361 ________. *Submarine.* viii + 212 pp., illus. (NavPers 16160-B) Issued with Perforations. Revision of *Fleet Type Submarine.* (For sale by Superintendent of Documents D208.11/2:Su1/2/961) LC 61-61258.

1362 ________. *Submarine Recognition Manual.* Washington, 1959. 38 pp., illus., diagrams. (NavPers 10011)

1363 ________. *Submarine Training Manuals.* Washington, 195- . LC 55-63571.

Some of the titles included in the series are:

The Fleet Type Submarine (NavPers 16160)
Submarine Main Propulsion Diesels. (NavPers 16161)
Submarine Electrical Installations. (NavPers 16162)
Submarine Refrigeration and A-C Systems. (NavPers 16163)
Submarine Distilling Systems. (NavPers 16163A)
Submarine Air Systems. (NavPers 16164)
Submarine Periscope Manual (NavPers 16165)
Submarine Trim and Drain Systems. (NavPers 16166)
Submarine Sonar Operator's Manual. (NavPers 16167)
Submarine Underwater Log. (NavPers 16168)

Submarine Hydraulic Systems. (NavPers 16169)

Torpedo Tubes, 21 Inch, Submerged, Mks, 32 to 39. (O.P. 1085)

1364 ________. *U.S. Submarine Losses World War II.* (Prepared by Commander, Submarine Force, Pacific Fleet, 1946) 1949 issue. ii+174 pp., illus. LC 49-45716.

1365 ________. *United States Submarine Losses. World War II.* 1949 issue. Washington, 1949. 174 pp., illus., ports. (NavPers 15, 784)

Includes sailing lists of crews and passengers of the 52 boats lost during WW II.

1366 USN. BUREAU OF ORDNANCE. *Torpedoes, Mark 15 Type.* (OP 642) Washington, 1945. 279 pp., illus., diagrams.

Technical manual for shipboard use. These torpedoes were primarily issued to destroyers, but with modifications could be used by submarines.

1367 USN. BUREAU OF SHIPS. *Diving Manual.* Washington, 1952. 160 pp., illus., diagrams. (Navships 250-880) LC 52-63204.

Supersedes the 1943 edition.

1368 ________. *Noise Survey and Repair Procedures for Submarine Noise Reduction.* Washington, 1951. vii, 80 pp., illus. (Navships 250-371). LC 51-61283.

1369 ________. *Submarine Material Guide.* Washington, 1944. 59 pp., illus., forms. LC 45-35297.

1370 ________. *Submarine Safety, Respiration and Rescue Devices.* Washington, 1946. 29 pp., illus. (Navships 250-000-94) LC 47-32849.

1371 ________. *Submarines.* Washington, 1947. 7 pp. (Navships 250-000-96). LC 47-32754.

1372 ________. *U.S. Navy Diving Manual.* Washington, 1956- . 1 volume, looseleaf, illus. (Navships 250-538) LC 57-61478.

Previous editions issued under the title: *Diving Manual.*

1373 ________. *U.S. Navy Diving Manual.* (D 211.6/2: D 64/970. S/N 0852-00012). Washington, 1970. 710 pp., illus.

Information on both "hard-hat" and Scuba diving.

1374 ________. *U.S. Navy Diving Manual: Change 1.* (D 211.6/2:D 64/970/Ch. 1. S/N 0852-00027). Washington, 1973. 191 pp., illus.

Revisions and additions to the manual up-dating the information.

1375 USN. COMMANDER OF SUBMARINES, PACIFIC. *United States Submarine Losses, World War II.* (Prepared by Commander Submarine Force, U.S. Pacific Fleet, 1946. For Official Use). Published in 1948. 176 pp., illus. LC 48-46922.

1376 U.S. NAVY DEPARTMENT. *Diving Manual, July 1916.* Washington, 1916. 122 pp., plates, plans, diagrams. LC 17-5390.

1377 ________. *Notes on the Over-Sea Towing of Destroyers and Submarines, May, 1913.* Washington, 1914. 16 pp., plates, diagrams. LC 14-30639.

U.S. Navy Diving Manual. See entries 1372, 1373, 1374, 1376.

1378 USN. FORTIFICATION BOARD. *Report of the Committee Upon Torpedoes Stationary and Movable, and Torpedo Boats.* Committee No. 2. Washington, 1885. 7 pp.

1379 USN. OFFICE OF NAVAL INTELLIGENCE. *Axis Submarine Manual.* (ONI#220-M) Washington, 1942. 34 pp., illus.

1380 ________. *A Brief Description of Submarine Mines. Warning! Avoid Them! Report All Discovered.* Washington, 1917. 8 numbered leaves. Illus., diagrams. LC 17-26372.

1381 USN. OFFICE OF NAVAL OPERATIONS. *ASW, Anti-Submarine Warfare.* (D207.2:An8) Washington, 1960. 38 pp., illus. LC 59-61216.

1382 ________. *Submarine in U.S. Navy.* (D207:2:Sul) Washington, Naval History Division, 1960. 32 pp., illus. LC 61-60130.

1383 ________. *War Instructions for United States Merchant Vessels.* August, 1918. 3rd ed. Washington, 1918. 165 pp., illus.. tables, forms. LC 42-28767.

When issued this was "Confidential." CSP 222. Addendum No. 1 to War Instructions for U.S. Merchant Vessels, CSP 206.

1384 USN. OFFICE OF NAVAL RECORDS AND LIBRARY. *American Ship Casualties of the World War Including Naval Vessels, Merchant Ships, Sailing Vessels, and Fishing Craft.*

Corrected to April 1, 1923. Washington, 1923. iii, 24 pp LC 23-26618.

1385 ________. *German Submarine Activities on the Atlantic Coast of the United States and Canada.* Washington, 1920. 163 pp., frontis., plates, charts, facsims. LC 20-27466.

1386 ________. *The Northern Barrage and Other Mining Activities.* Washington, 1920. 146 pp., port, plates, charts, tables, diagrams. LC 20-27467.

1387 USN. OFFICE OF NAVAL RESEARCH. REPORTS. #5182. *Algae and Submarine Habitability.* By John M. Leonard. August 26, 1958. 11 pp., bibliog.

1388 ________. #5465. *Present Status of Chemical Research in Atmospheric Purification and Control on Nuclear Powered Submarines.* Edited by R. R. Miller and V. R. Piatt. April 21, 1960. 173 pp., bibliog.

1389 ________. #5545. *Shock Design of Shipboard Equipment. Part I. Dynamic Design-Analysis Method.* By R. O. Belsheim and G. J. O'Hara. September 16, 1960. 60 pp., illus., bibliog.

1390 ________. #5563. *Dry Chemical System for Carbon Dioxide Removal from Submarines.* By S. H. Smith, Jr. November 23, 1960. 10 pp., illus., bibliog.

1391 ________. #5609. *Procedures for Evaluation of Fracture Toughness of Pressure-Vessel Materials.* By W. S. Pellini and J. E. Srawley. June 8, 1961. 30 pp., illus., bibliog.

1392 ________. *Annual Report on the Present Status of Chemical Research in Atmospheric Purification and Control on Nuclear Powered Submarines.* Edited by V. R. Piatt and E. A. Ramskill. July 14, 1961. 142 pp., illus., bibliog.

1393 ________. *Project Sealab Summary Report. An Experimental Eleven-Day Undersea Saturation Dive at 193 Feet.* Sealab I Project Group: H. A. O'Neal, G. F. Bond, R. E. Lanphear, and T. Odum. (ONR Report ACR-108) Washington, 1965. 60 pp., illus.

1394 USN. SUBMARINE BOARD. *Report and Recommendations on Submarine Safety and Salvage. . . .* April 1, 1929. Washington, 1929. iii, 8 pp., tables. LC 29-26413.

1395 USN. SUBMARINE SCHOOL. NEW LONDON. *Basic Enlisted Submarine Text.* Part I. (Navpers 10490) Washington, 1963. 215 pp., illus., diagrams.

Revised version of the earlier text *The Submarine* (Navpers 16160)

The United States Nuclear Navy. See entry 489.

1396 U.S. PRESIDENT (Wilson). *Relations With the German Government.* Address of the President of the United States, Delivered at a Joint Session of the Two Houses of Congress, April 19, 1916. Washington, 1916. 6 pp. (64th Congress, 1st Session. House. Document 1034). LC 16-26400.

1397 U.S. PRESIDENT (Wilson). *Address of the President of the United States, Delivered at a Joint Session of the Two Houses of Congress, February 3, 1917.* Washington, 1917. 6 pp. LC 17-26239.

1398 ________. *Address of the President of the United States Delivered at a Joint Session of the Two Houses of Congress, February 26, 1917.* Washington, 1917. 6 pp. LC 17-26240.

1399 U.S. PRESIDENT, 1921-1923. (Harding). *Conference on the Limitation of Armament.* Address of the President of the United States Submitting the Treaties and Resolutions Approved and Adopted by the Conference . . . together with the Report of the American Delegation . . . February 9, 1922. Washington, 1922. xiii, 132 pp. (67th Congress. 2nd Session. Senate. Document 125). LC 22-26125.

1400 U.S. SHIPPING BOARD. *Instructions Which Must Be Followed by Licensed Officers of all United States Merchant Vessels Traversing the Submarine Zone.* Washington, 1917. 8 pp. LC 19-26266.

1401 ________. *Suggested Regulations for the Protection of Ships Imperiled by Mines or Submarines.* Washington, 1917. 13 pp. LC 19-9299.

USS Mudskipper. A Novel of World War II. See entry 567.

USS Seawolf: Submarine Raider of the Pacific. See entry 429.

USS Triton, SSRN 586, First Submerged Navigation. See entry 75.

U.S. Submarine Base, New London, Conn. See entry 57.

V

Vacances Chez Neptune. See entry 1181.

Vahan, Richard. See entry 348.

1402 VALENTINER, Max, 1883- . *Der Schrecken der Meere: Meine U-Boot-Abenteuer.* Zurich, Amalthea-Verlag, 1931. 327 pp., illus., ports. LC 49-44151.

Memoirs of WW I U-Boat ace.

1403 ________. *La Terreur des Mers: Mes Aventures en Sous-Marin, 1914-1918.* Translated by P. Teillac. Paris. Payot, 1931. 252 pp. LC 32-2365.

1404 ________. *Il Terrore sui Mari: le Avventure di un Commandante di Sommergibili.* Translated by Renzo Segala. 2nd edition. Milan, Fratelli Treves, 1936. 277 pp., frontis. port, plates. LC 45-53635.

1405 ________. *300000 Tonnen Versenkt! Meine U-Boots-Fahrten.* Berlin and Vienna, Ullstein, 1917. 154 pp., plates, ports. LC 20-11296.

1406 ________. *U38: Wikingerfahrten eines Deutschen U-Bootes.* Berlin, Ullstein, 1934. 269 pp., plates, ports. LC 34-22563.

1407 VANDERKOGEL, Albert. *Underwater Sport,* with Rex Lardner. Line drawings by Paul Norris. NY, Holt, 1955. 188 pp., illus. LC 55-7919.

VanderKogel, Albert. See also entry 1018.

Van Doren, Carl. See entry 641.

Van Drebbel, Cornelius. See Drebbel, Cornelis, 1572-1633.

Van Tromp, Cornelius. See entry 1455.

Varekrieg. See entry 924.

Vasa (ship). See entry 435.

Vastness of the Sea. See entry 499.

Vatine, Comte de la. See entry 998.

Vaughn, Frank. See entry 1438.

1408 VEDEL, Émile Henry Auguste, 1858- . *Quartre Années de Guerre Sous-Marine.* Paris, Plon-Nourrit, 1919. xvi, 366 pp. LC 22-7221.

Veertig Jaren Onderzeedienst 1906-1946. See entry 241.

Vengeance in the Depths. See entry 1174.

Verdammter Atlantik. . . . See entry 594.

La Vérité sur la Guerre Sous-Marine. See entry 873.

1409 VERNE, Jules, 1828-1905. *Twenty Thousand Leagues Under the Sea.* Paris, 1870.

This classic of imaginative submarining by the "Father of Science Fiction" has been reprinted many times, Including an edition published by George M. Smith in Boston in 1875. Othei editions include: Associated Booksellers, Dodd (Great Illustrated Classics series), Doubleday (Dolphin), 1960, Dutton (Everyman's), Globe, 1955, Grosset, 1954 and 1958, Norton (Collins Classics) 1959, Rand McNally, 1954, Scott (adapted) 1955, World (Rainbow Classics).

NY, Heritage Press, 1956, has 325 pp. with colored illustrations by Edward A. Wilson. The translation is by Mercier Lewis, and there is an introduction by Fletcher Pratt.

A definitive modern translation by Walter James Miller, N.Y. University assisted by Judith Ann Tirsch. Afterword by Damon Knight. Illustrated by Walter Brooks. (NY, Washington Square Press, 1965. 389 pp., illus. LC 65-25245)

London, J. M. Dent and NY, E. P. Dutton, 1966. 305 pp., illustrated by William McLaren.

Santa Rosa, Calif., Classic Press Inc., 1968. Illustrated by Don Irwin. Large format. Text pp. 1-274. A "Backword" between pp. 276-277, concerning Jules Vernes and including a portrait. "Prepare to Dive! The Story of the Submarine" between pp. 278-283, which is an illustrated epitome history of submarines.

1410 ________. *Vingt Milles Lieues Sous les Mers.* Paris, J. Hetzel et cie., 1871. 434 pp., illus. LC 2-6337.

The illustrations are by Edouard Riou and Alphonse de Neuville.

Die Verratene Flotte. See entry 442.

Versunkene Schätze. See entry 550.

Die Verwendung von Minen im Seekrieg. See entry 1168.

Vessels for Underwater Exploration. See entry 780.

1411 *Vessels Sunk by German Submarines, Mines, or Warships.* Data concerning the sinking of neutral vessels belonging to Norway, Sweden, Denmark, and Holland, which were sunk . . . between the dates of August 1, 1914 and March 25, 1916. Washington, 1916. 6 pp. (64th Congress, 1st Session. Senate. Doc. 381.) LC 16-26330.

Vi Som Var Våbenløse. See entry 944.

Vicar, Henry, pseud. See Felsen, Gregor.

Vickers, Ltd. See entry 1143.

Victoria (ship). See entries 304, 305.

The Victoria and the Triton. See entries 304, 305.

Victory at Sea. See entry 950.

The Victory at Sea. See entry 1136.

Vie du Pere Mersenne. See entry 295.

La Vie et les Ouvrages de Denis Papin. See entry 1089.

1412 VILLANÚA, León, 1818- . *Peral, Marino de España: Biografiá.* Madrid, Colección Europa, 1934. 251 pp., illus., facsims, ports. LC 34-38988.

1413 VILLENEUVE-TRANS, Romée François, comte de, 1888- . *La Liberté des Mers. Le Blocus de L'Allemagne, La Guerre Sous-Marine.* Paris, Pedone, 1917. 101 pp., map. LC 18-9441.

1414 VILLON, A. M., 1863-1895. *La Navigation Sous-Marine.* Paris. Tignol, 1890. 84 pp., illus., diagrams.

Vindictive, HMS (cruiser). See entry 56.

Vingt Milles Lieues Sous les Mers. See entry 1410.

Vision. See entries 1341, 1342.

Visual Characteristics of Submarine Population. See entry 1341.

Vita Segreta de Sommergibili. See entry 273.

Vitale, Vincent. See entry 1101.

Vodolaznoe Snariazhenie. See entry 1130.

Voenno-Morskoi Flot. See entries 719, 906.

Voenno-Morskoi Flot Sovetskoi. . . . See entry 725.

1415 VOGEL, Hans, 1852- . *Das Moderne Unterseeboot.* Berlin, Springer, 1911. 20 pp.

A reprint of an article from *Zeitschrift des Vereines Deutscher Ingenieure.* Jahr. 1911.

Vogel, Joseph. See entry 1349.

Vom U-Boot zur Kanzel. See entry 918.

1416 VON der PORTEN, Edward P. *The German Navy in World War II.* London?, Apollo Editions. 1972. 274 pp., illus., maps.

1417 ________. *The German Navy in World War Two.* Foreword by Karl Doenitz. London?, Pan Books, 1972. 286 pp., illus. ISBN 0-330-23235-5.

Vore Undervandsbåde Gennem 50 Dr. See entry 1176.

Voyage of the Deutschland. See entry 717.

Voyage to the Bottom, 1955-1965. See entry 123.

Voyages. See entries 71, 75, 304, 305.

1418 VRIES, W. P. de. *De Duikboot en Haar Rol in den Europeeschen Oorlog.* Amsterdam, Graauw, 1915. 82 pp., plates.

W

Waffe Unter Wasser. See entry 1460.

1419 WAGNER, Frederick, 1928- . *Famous Underwater Adventures.* NY, Dodd, Mead, 1962. 159 pp., illus. LC 62-8750.

"An enjoyable collection of biographical sketches of 13 men—scientists, inventors, divers, submariners, and salvage workers—who through their achievements in the underwater world, have enabled man to explore the sea and to make use of it. Presented are Edmond Halley, Robert Fulton, Wilhelm Bauer, Simon Lake, William Beebe, Auguste Piccard, Jacques Piccard, Edward Ellsberg, C. B. Momsen, Maurice Ewing, Jacques-Yves Cousteau, Edward L. Beach, and Hannes Keller. A few photographs are included and a five-page list of adult and children's books for further reading is appended. Grade 7 and up."—from *The Booklist.*

1420 ________. *Submarine Fighter of the American Revolution: The Story of David Bushnell.* NY, Dodd, 1963. 145 pp., plates. LC 63-8711.

The prepublication title was announced as *David Bushnell and the American Revolution.*

Wahoo, USS (SS 238). See entry 1185.

Wake of the Wahoo. See entry 1185.

1421 WALDEYER-HARTZ, Hugo von, 1876- . *U-Boot und U-Boot-Krieg.* Braunschweig, Westermann, 1917. 252 pp., plates, port. LC 51-50705.

1422 WALDRON, Thomas John. *The Frogmen: The Story of the Wartime Underwater Operators.* with James Gleeson. London, Evans, 1950. 191 pp., illus., ports. LC 51-751.

Wales, Douglas. See entry 552.

Walker, Art. See entry 1142.

Walker, R. N. See entry 1043.

1423 WALKER, Sydney Ferris. *Submarine Engineering: All About Work Under Water. . . .* London, Pearson, 1914. 126 pp., plates, diagrams. LC A 15-885.

Walter U-Boote. See entry 730.

1424 WALTERS, E. W. *Submarine Stories and Adventures: With an Introduction on the Construction and Use of German Submarines, Torpedoes, and Mines.* London, Kelly, 1917. 256 pp., frontis.

Waltrip, Mildred. See entry 652.

The War at Sea. See entry 1063.

War Beneath the Sea. See entries 125, 154.

War in the Underseas. See entry 1449.

War Instructions for U.S. Merchant Vessels. See entry 1383.

The War on German Submarines. See entry 198.

War on the Sea: Strategy and Tactics. See entry 285a.

War Under the Waves. See entry 1430.

Warfish. See entries 527, 528.

Warhead. See entry 481.

1425 WARREN, Charles Esme Thornton and James BENSON. *Above Us the Waves.* (edited for schools by R. T. C. Hughes) London, Harrap, 1964. 176 pp., maps.

1426 ________. *The Admiralty Regrets . . . The Story of HMS Submarine Thetis.* London, Harrap, 1958.

Published in USA under the title *Only Four Escaped.*

1427 ________. *The Midget Raiders: The Wartime Story of Human Torpedoes and Midget Submarines.* NY, Sloane, 1954. 318 pp., illus. LC 53-9342.

Published in Britain under the title of *Above Us the Waves.*

1428 ________. *Only Four Escaped: The Sinking of the Submarine Thetis.* NY, Sloane, 1959. 219 pp., illus. LC 59-7411.

1429 ________. *Will Not We Fear: The Story of His Majesty's Submarine Seal and of LCDR Rupert Lonsdale.* Foreword by R. P. Lonsdale. London, Harrap, 1961. Toronto, Clarke, Irwin, 1961. NY, Sloane, 1962. 228 pp., illus., plates. LC 62-6795.

War's New Weapons. See entry 303.

Warship Vasa, Deep Diving and Marine Archaeology in Stockholm. See entry 435.

Warships in Profile. See entries 996, 1476.

1430 WARSHOFSKY, Fred. *War Under the Waves.* NY, Pyramid Books, 1962. 157 pp.

Stories of submarine warfare from Turtle through World War Two as interpreted by the author.

Was Jeder vom Deutschen U-Boot Wissen Muss. See entry 55.

1431 WASHINGTON, D.C. CONFERENCE ON THE LIMITATION OF ARMAMENT, 1921-1922. *Armament Conference Treaties.* Treaties and Resolutions approved and adopted by the Conference on the limitation of armament submitted by the President of the United States to the Senate for advice and consent to their ratification. . . . Washington, 1922. 44 pp. (67th Congress. 2nd Session. Senate. Doc. 124) LC 22-26124.

1432 ________. *Conference on Limitation of Armament. Washington, 1921-22.* London, HMSO, 1922. 88 pp. (Gt. Brit. Foreign Office. Miscellaneous No. 1. 1922) LC 23-3483.

1433 ________. *Conference on the Limitation of Armament. . . .* Washington, 1922. 1757 pp. English and French. LC 22-27501.

1434 ________. *Conferencia Sobre Limitación de Armamentos. . . .* NY, Asociación Americana para la Conciliación Internacional, 1922. 221 pp. LC 24-11773.

1435 ________. *La France et les Sous-Marins. Réponse de M. Jusserand, Ambassadeur de France, à Lord Lee de Fareham, Premier Lord de L'Amirauté . . .* Washington, 1922. 818-829 pp. LC 23-26357.

1436 ________. *Treaties and Resolutions of the Conference on the Limitation of Armament, As Ratified by the United States Senate; Facts and Tables.* NY and Washington, Federal Trade Information Service, 1922. 60 pp., tables. LC 22-7852.

1437 ________. *Washington Conference on the Limitation of Armaments.* NY and Greenwich, Conn., American Association for International Conciliation, 1921-22. xxxiii, 382 pp. LC 21-22100.

Watchdogs of the Deep. See entry 689.

Watkins, K. See entry 115.

We Captured A U-Boat. See entry 468.

We Clear the Way. See entry 391.

We Dive at Dawn. See entries 358, 359.

We Were There on the Nautilus. See entry 1438.

We Work for Freedom. See entry 625.

1438 WEBB, Robert N., 1906- . *We Were There on the Nautilus.* Illustrated by Frank Vaughn. NY, Grosset and Dunlap, 1961. 178 pp., illus. ISBN 0-448-05035-8. LC 61-66511.

For audience of grades 4 through 9. William R. Anderson served as consultant to the author.

Webfooted Warriors. See entry 609.

1439 WEDDIGEN, Otto. *Das Erste Handels-Unterseeboot Deutschland und Sein Kapitan Paul Koenig.* Leipzig, Post & Obermueller, 1916. 157 pp.

1440 ________. *Unser Seehelden Weddigen, Leben und Taten.* Berlin, Scherl, 1917. 85 pp.

1441 WEDDIGEN, Otto Eduard, 1880-1915. *Otto Weddigen und Seine Waffe, aus Seinen Tagebuchern und Nachgelassenen Papieren Unter Mitwirkung der Familie.* Edited by Hermann Kirchhoff. Berlin, Marinedank, 1915. 157 pp., frontis., illus., plates, ports. LC 16-16532. Rev.

Weddigen, Otto Eduard, 1880-1915. Commanding U-9 on 9/22/1914 sank the three British cruisers Aboukir, Hogue, and Cressey in a period of 135 minutes. See also entries 705, 1034, 1440.

Wegduiken: De Nederlandse Onderzeedienst, 1906-1966. See entry 1188.

Weise, Gerhart. See entry 573.

1442 WEISS, Malcolm E. *Man Explores the Sea.* Maps and diagrams by Lloyd Birmingham. NY, Messner, 1969. 110 pp., illus. SBN 671-32202-8 and 671-32203-6. LC 89549.

Oceanography written for grades 4 through 6 audience. Covers three Sealabs, and the Ben Franklin Gulf Stream drift voyage.

Weiss, Sylvia E. See entry 1323.

1443 WELLER, George Anthony, 1907- . *The Story of Submarines.* (Landmark Book) NY, Random House, 1962. 210 pp., illus. LC 62-9686.

Weller was awarded the Pulitzer Prize in Journalism in 1943 for his story of an emergency appendectomy which was performed in a submerged submarine in the war zone. He also wrote under the pseudonym of Michael Wharf.

Wells, Robert. See entry 1037.

Welsh, Vernon. See entry 473.

Weltkriegsbücherei. See Zeitgeschichte-Weltkriegsbücherei.

Werewolf der Meere. See entry 882.

1444 WERNER, Herbert A. *Iron Coffins: A Personal Account of the German U-Boat Battles of World War II.* NY, Holt, Rinehart & Winston, 1969. 329 pp., illus. SBN 03-081322-0. LC 71-80336.

1445 WERTH, Alexander, 1901- . *Russia at War: 1941-1945.* London, Barrie and Rockliff, 1964. NY, Dutton, 1964. Toronto, Smithers, 1964. xxv, 1100 pp., maps. LC 64-19533.

Werts, Margaret F. See entry 1277.

1446 WESTERMAN, P. F. *Pirate Submarine.* Toronto, Musson, 1945. 251 pp.
Fiction.

1447 WESTWOOD, J. N. *The Illustrated History of the Russo-Japanese War.* Chicago, Regnery, 1973. 126 pp., illus.

Weybrew, Benjamin B. See entries 1343, 1347, 1349, 1350, 1353.

What a Wonderful Machine Is the Submarine. See entry 61.

What the Citizen Should Know About Submarine Warfare. See entry 1479.

Wheeler, Ardis Hillman. See entry 245.

1448 WHEELER, Harold Felix Baker, 1877- . *Daring Deeds of Merchant Seamen in the Great War.* London, Harrap, 1918. 319 pp., frontis., plates. LC 19-3022.

1449 ________. *War in the Underseas.* NY, Crowell, 1919. 319 pp., frontis., plates. LC 20-7152.

When the U-Boats Came to America. See entry 224.

1450 WHITAKER, Herman, 1867-1919. *Hunting the German Shark: The American Navy in the Underseas War.* NY, Century, 1918. 310 pp., frontis., illus., port, maps, plates. LC 18-23106.

White, A. O. See entry 1282.

1451 WHITE, Robb. *Up Periscope.* Garden City, Doubleday, 1956. 251 pp. LC 56-9060.

1452 ________. *Up Periscope.* NY, Scholastic Book Services, 1960. 220 pp.

1453 WHITEHOUSE, Arch, *pseud. Subs and Submariners.* NY, Doubleday, 1961. Toronto, 1961. 335 pp., plates. LC 61-8909.

Same title. London, Muller, 1963. 416 pp., plates.

Author's full name is Arthur George Joseph Whitehouse, 1895- .

1454 WHITESTONE, Nicholas. *The Submarine: The Ultimate Weapon.* London, David-Poynter, 1973. 146 pp. LC 73-173627.

Former RN officer and ex-naval correspondent traces the evolution of the submarine. Its influence on world politics and on offensive and defensive strategies.

Who Really Invented the Submarine? See entry 1485.

1455 *Who Won the Great War? The Revelations of the Late Cornelius Van Tromp Turned the Tide of the Great War, The Second Day of May 1917, and Made Possible the Victory at Sea and the Victory on Land.* West Hoboken, N.J., Weaver, 1921. 31 pp., illus., ports. LC 43-43037.

Material relating to William Russell Hinsdale and his convoy system.

1456 WHYTE, Adam Gowans, 1875- . *Deep-Sea Salvage,* with Robert L. Hadfield. London, Low, Marston & Co., 1933. xi, 244 pp. plates. LC 34-20774.

1457 ________. *Divers and Diving.* London, Pitman, 1938. 128 pp., illus. LC 39-2075.

1458 WIDDER, Arthur. *Action In Submarines.* NY, Harper & Row, 1967. 213 pp., illus. ISBN 0-06-026403-9. LC 66-18650.

Journalistic survey of submarine history written for juvenile audience.

1459 WIEBICKE, Karl. *Die Männer von U96: Erinnerungen an Fahrten Unseres U-Bootes.* Leipzig, Koehler, 1934. 208 pp., illus., maps, plates, ports. LC 35-4129.

1460 WIEDEMEYER, Gerhard. *Waffe Unter Wasser, Die Erfindung des Unterseebootes und seine Geschichte.* Berlin,

Kyffhauser, 1940. 160 pp., illus., ports. LC AF 48-4156.

1461 WIGUM, Harald Gabriel Ueland, 1890- . *U-Baatofrene: Billeder fra Tysklands Krig Mot Vore Sjøfolk.* Bergen, Tønnessen, 1917. 92 pp. LC 21-16136.

Wilhelm Bauer, der Erfinder des Unabhangigen Unterseeboots. See entry 493.

1462 WILKES, Bill St. John. *Nautical Archaeology. A Handbook for Skin Divers.* NY, Stein and Day, 1971. 294 pp., illus., diagrams, bibliog. ISBN 0-8128-1415-0. LC 77-159558.

1463 WILKINS, Sir George Hubert, 1888-1958. *Under the North Pole: The Wilkins-Ellsworth Submarine Expedition.* NY, Brewer, Warren & Putnam, 1931. London, Benn, 1931. xiv, 347 pp., frontis., illus., map, plates, ports, facsims. LC 31-11064.

Wilkins and his group purchased an O-Class submarine from the U.S. Navy and converted it for Arctic operations and renamed it the Nautilus. The book was issued in at least three editions: a regular trade edition, a leather-bound "De Luxe" edition which sold for $50.00, and a special "Contributor's Edition" which was contained in a stainless steel box, at $600.

1464 WILKINS, John, Bishop of Chester, 1614-1672. *Dissertations on Submarine Navigation as a Result of His Studies.* London, 1848.

1465 ________. *Mathematical Magick.* London, 1648 and 1680.

Wilkins-Ellsworth Submarine Expedition. See entry 1463.

1466 WILKINSON, Burke, 1913- . *By Sea and By Stealth.* NY, Coward, McCann, 1956. Toronto, Longmans, 1956. 218 pp., illus. LC 56-11525.

Wilkinson, CDR Eugene. See entry 897.

Will Not We Fear. See entry 1429.

1467 WILLIAMS, G. *Diving for Treasure and Other Adventures Beneath the Sea.* London, Faber and Gwyer, 1926. 157 pp., illus., plates, port. LC 27-6972.

1468 WILLIAMS, E. Edwin. *The Submarine Force Library and Museum.* A Thesis submitted to the faculty of the State University of New York College at Oneonta at its Cooperstown Graduate Programs for the Degree of Master of Arts. 1971 192 pp., illus., appendices, map.

Researched and written by the former director of the library.

1469 WILLIAMS, Wirt. *The Enemy.* (Signet Book) NY, New American Library, 1956. (Originally published NY, Houghton Mifflin, 1951)

Fiction, ASW during the WW II Battle of the Atlantic.

1470 WILLIAMSON, Charles, 1855- . *Under the Sea via the Williamson Flexible Tube System.* Norfolk, Va., Williamson Submarine Corp., 1915. 67 pp., illus. LC 15-11703.

Williamson, Jack. See entries 980-983.

Williamson Flexible Tube. See entry 1470.

Willm, P. H. See entry 631.

1471 WILSON, Eugene Edward, 1887- . *Kitty Hawk to Sputnik to Polaris: A Contemporary Account of the Struggle Over Military and Commercial Air Policy in the United States.* Palm Beach, Fla., Lit. Investors Guild, 1967. 231 pp., illus. LC 67-4372.

1472 WILSON, Herbert Wrigley, 1866- . *Hush; or, the Hyrdophone Service.* London, Mills & Boon, 1920. vii, 188 pp., frontis., plates, ports., diagrams. LC 21-3795.

Wilson, L.P.R. See entries 168, 169.

Windows in the Sea. See entries 783, 1119.

1473 WINGATE, John. *Jimmy-the-One: A Submariner Sinclair Story.* London, Newnes, 1960. 256 pp.

WW II fiction.

1474 ________. *Jimmy-the-One: A Submariner Sinclair Story.* Toronto, British Book Service, 1960. 256 pp.

1475 ________. *Submariner Sinclair.* London, Newnes, 1959. Toronto, British Book Service, 1959. 256 pp. LC 60-17803.

1476 WINGATE, John, ed. *Warships in Profile, Vol. I.* NY, Doubleday, 1972. *Volume II* NY, Doubleday, 1973.

Each volume contains twelve chapters each dealing with individual ships.

1477 Winton, John, pseud. *Down the Hatch.* London, M.

Joseph, 1961. Toronto, Collins, 1961. NY, St. Martin's Press, 1962. 204 pp. LC 62-8324.

Humorous fiction about the HMS Seahorse. Author also wrote *We joined the Navy* (1959) and *We Saw the Sea*. (1960).

Wir an den Maschinen. See entry 1024.

Wir Jagen Deutsche U-Boote. See entry 185.

Wir Jagten Panzerkreuzer. . . . See entry 1163.

Wir Leben Noch! See entries 904, 905.

Wire-Roping the German Submarine. See entry 1053.

With Blood and Iron. See entry 1020.

With the Red Fleet. See entry 494.

Der Wolf im Atlantik. See entry 1048.

1478 WOLFE, Louis. *Ships That Explore the Deep*. NY, Putnam, 1971. 94 pp., illus. LC 75-125324.

For grades 6 through 9.

Wolfpack. See entry 568.

Die Wölfe und der Admiral. See entry 434.

The Wonders of the Submarine. See entry 255.

1479 WOODBURY, David Oakes, 1896- . *What the Citizen Should Know About Submarine Warfare*. Drawings by Louis H. Ruyl. NY, Norton, 1942. 231 pp., illus., diagrams, bibliog. LC 42-20590.

Later reprints bore the title *What You Should Know About Submarine Warfare*.

1480 WOODWARD, David, 1909- . *The Russians at Sea: A History of the Russian Navy*. NY, Praeger, 1966. 254 pp., maps. LC 66-12477.

Also pub. without the subtitle in London by Kimber in 1965. 254 pp., maps. LC 65-6332.

1481 ________. *The Secret Raiders: The Story of the German Armed Merchant Raiders of the Second World War*. London, 1955. NY, Norton, 1955. 288 pp., plates. LC 55-14915.

Woodward, David. See also entry 318.

The World Beneath the Sea. See entries 54, and 347.

The World Beneath the Waves. See entry 334.

1482 WORLD CONGRESS OF UNDERWATER ACTIVITIES. Second. London, 1962. *The Undersea Challenge.* A report of the proceedings of the Congress, which was organized by the British Sub-Aqua Club for the Confederation Mondiale des Activities Subaquatiques. Edited by Bernard Eaton. Kingston-on-Thames, Surrey, England, British Sub-Aqua Club, 1963. 182 pp., illus., maps.

World Underwater Federation. See entry 1219.

World War One, 1914-1918. See entries, 1, 7, 9, 11, 17, 30, 31, 33, 37, 39, 41, 56. 109, 110, 120, 126, 135, 142, 160, 164, 165, 173, 182, 183, 184, 185, 187, 190-194, 198. 201, 202, 206, 210, 213, 214, 216-219, 222, 223, 224, 246, 264, 276, 277, 297, 298. 307, 315, 325, 327, 331, 332, 335, 358, 359, 366, 394, 395. 400, 401, 415-421, 427, 428, 438-442, 446-449, 460, 461, 469, 471, 477-480, 484, 488, 490. 502, 505-509, 513-516, 521, 531, 535 536, 543, 545, 548, 549, 559, 576, 577, 585, 591, 598. 605, 608, 610, 614, 615, 620, 624, 636, 644-648, 660, 664, 667, 671, 672, 679, 680-684, 689, 692. 701-704, 706, 707, 711, 713, 717, 727, 733, 734, 737, 738, 747, 748, 750, 753, 756, 757, 758, 763, 775, 786. 804, 805, 837, 845, 846, 857, 859, 860, 865, 870, 871, 873, 882, 891. 900, 902, 904, 905, 909, 910, 911, 915-918, 924-929, 931, 942, 944, 947, 953, 968, 976, 977, 978, 984. 1011, 1023, 1024, 1034, 1038, 1053, 1067, 1077, 1087, 1096, 1097. 1104, 1105, 1107, 1108, 1109, 1113, 1114, 1115, 1135, 1136, 1143, 1156-1163, 1165-1167, 1178, 1189, 1194. 1228, 1232-1235, 1239, 1240, 1261, 1262, 1263, 1273, 1274, 1284. 1315, 1316, 1321, 1324, 1383-1386, 1396, 1397, 1398. 1400-1406, 1408, 1411, 1413, 1418, 1421, 1440, 1441, 1448, 1449, 1450, 1455, 1459, 1461, 1472.

World War Two., 1939-1945. See entries: 12, 13, 45, 46, 52, 55, 62-70, 72, 73, 74, 77, 79, 80, 85-91, 94. 103, 125, 130, 139, 144, 153-156, 159, 166-169, 171, 189, 195, 199. 203, 207, 208, 209, 230, 231, 251, 252, 253, 274. 314, 342, 354, 379, 389, 390, 391, 392, 396. 407, 429-434, 434, 436, 444, 466, 467, 468, 482, 486, 492. 510, 511, 512, 517, 527, 528, 538, 544, 545, 548, 551, 552, 553, 562-568, 571, 572, 573, 578, 579, 580, 583, 584, 594. 601, 602, 604, 605, 609, 618, 621, 630, 634, 637-642, 654, 661, 665, 686, 693, 697. 700, 770-773, 779, 790-796. 806, 808, 814, 819, 822, 823, 826, 832, 839, 850-853, 856, 858, 884.

912, 923, 933, 934, 936, 941, 943, 950, 955, 956, 957, 974, 998, 999. 1019, 1020, 1041-1048, 1054, 1055, 1060, 1062, 1063, 1065, 1069, 1070, 1071, 1073, 1090, 1091, 1091a 1103, 1110, 1111, 1112, 1126, 1127, 1150, 1152, 1185, 1193. 1220, 1231, 1238, 1257, 1266, 1268. 1326, 1364, 1365, 1375. 1416, 1417, 1422, 1425, 1427, 1429, 1444, 1445, 1460, 1469, 1481, 1486, 1487, 1489, 1490.

World Without Sun. See entry 262.

The Wreck Hunters. See entry 678.

1483 WRIGHT, Helen, 1914- and S. B. RAPPORT, editors. *Great Undersea Adventures.* NY, Harper, 1966. 381 pp. LC 66-18652.

Juvenile.

Wright, Jerauld. See entries 777, 1047.

1484 WURZER, D. F. *De Papino et Machina Papiniana.* Marbourg, 1809.

1485 WYCKOFF, James. *Who Really Invented the Submarine?* NY, Putnam, 1965. Don Mills, Longmans, 1965. 95 pp., illus. LC 65-10860.

Juvenile.

Wytema, M.S. See entry 781.

X·Y·Z

The X-Craft Raid. See entry 466.

The Yankee-Mining Squadron. See entry 91.

1486 YOKOTA, Yutaka with Joseph D. HARRINGTON. *The Kaiten Weapon.* (S600) NY. Ballantine Books, 1962. 256 pp., illus.

1487 ________. *Suicide Submarines.* (U6065) NY, Ballantine Books, 1962. 255 pp., illus.

The Japanese "Kaiten" one or two man submarines, were human torpedoes employed during WW II.

Young, Edward. See entry 468.

1488 YOUNG, Edward Preston, 1913- . *Look At Submarines.* Illustrated by John Robinson. London, Hamilton, 1964 95 pp., illus.

Juvenile.

1489 ________. *One of Our Submarines.* London, Hart-Davis, 1952. Toronto, Clarke, Irwin, 1952. 316 pp., plates, maps. LC 53-1105.

Published in USA as *Undersea Patrol.*

1490 ________. *Undersea Patrol.* NY, McGraw-Hill, 1953. 298 pp., plates, maps. LC 53-10631.

The Young Naval Captain. See entry 124.

The Young Sportsman's Guide to Scuba Diving. See entry 368.

Youniss, Richard P. See entry 1352.

Zeebruge. . . . See entry 976.

Zeebrügge. Der Englische Angriff auf die Deutsche U-Boot-Basis. See entry 56.

Zeebrügge-Ostend Raids, 1918. See entry 56.

Zehn Jahre und Zwanzig Tage. See entry 320.

1491 ZEITGESCHICHTE-WELTKRIEGSBÜCHEREI. (Modern History-World War Library) Stuttgart, Germany. *Classified Catalog.* 20 volumes. *Alphabetical Catalog.* 11 volumes. Boston, G. K. Hall & Co., 1968.

These bibliographical reference books are available in major research libraries. The Weltkriegsbücherei is a major, specialized research library concerning war and the causes of war in the 20th century. It is similar in scope to the American Hoover Library at Stanford University. The Director, Jürgen Rohwer (q.v.) has written several books about German submarines.

1492 ZHIGALOV, Ivan Matveevich, comp. *Podvodniki.* Moscow, 1956. 261 pp., illus. LC 57-15787.

1493 ZIM, Herbert Spencer, 1909- . *Submarines, The Story of Undersea Boats.* Illustrated with drawings by James MacDonald. NY, Harcourt, Brace, 1942. xii, 306 pp., frontis, illus., diagrams. LC 42-19801.

1494 ZOLOTOVSKII, Konstantin Dmitrievich. *Deep Sea Divers.* Translated by Beatrice Kinkead. Illustrated by H. Shevlachov. NY & Phila., Lippincott, 1938. 212 pp., frontis., plates. LC 38-12381.

An account of the author's experiences as a deep sea diver.

1495 ————. *Podvodnye Mastera.* Moscow, 1950. 193 pp., illus. LC 52-47244.

Zoomies, Subs and Zeroes. See entry 794.

1496 ZUKOWSKI, Olgierd. *Stalowe Rekiny, Krotka Historja i Opis Lodzi Podwodnych.* Warsaw, Wojskowy Instytut, 1931. 87 pp., illus., plates, map, bibliog. LC CA 32-576 Unrev'd.

Zumwalt, Elmo R., Jr. See entry 498.

THE END